Peterson's
egghead's Guide to Algebra

Cara Cantarella

About Peterson's

Peterson's provides the accurate, dependable, high-quality education content and guidance you need to succeed. No matter where you are on your academic or professional path, you can rely on Peterson's print and digital publications for the most up-to-date education exploration data, expert test-prep tools, and top-notch career success resources—everything you need to achieve your goals.

For more information, contact Peterson's, 800-338-3282 Ext. 54229; or find us online at www.petersonsbooks.com.

ISBN-13: 978-0-7689-3778-7

Printed in the United States

10 9 8 7 6 5 4 3 2 1   15 14 13

Contents

Contents

Contents

Chapter 7: Polynomial Equations. 139

Chapter 8: Absolute Value 171

Chapter 9: Inequalities. 185

Contents

Before You Begin

Welcome to *egghead's Guide to Algebra*! My name is egghead, and I'll be your guide throughout the book.

Before You Begin

This egghead's Guide was designed to help you learn algebra in a fun and easy way. Sometimes learning can be . . . well, boring. It can also be confusing at times. If it wasn't, we'd all have straight A's, right?

As your guide through the adventure of education, I'm here to make things a bit more enjoyable. I studied the boring books so you don't have to. I got straight A's and lived to tell about it. I understand this stuff, and you can too. In this guide, I'll show you what you need to learn to get to the next level.

Wherever I can, I explain things in pictures and stories. I break concepts down and teach them step by step. I try to stick with words that you know. I give examples from real life that you can relate to.

I want you to succeed, and I know you can do it!

In this book, we'll work together to improve your algebra skills and build your confidence. Confidence is very important, and it comes from trust. You can trust me as your guide, and most important, you can trust yourself. If your algebra knowledge isn't strong enough, let's do something about it!

How this book is organized

This book contains eleven chapters. We recommend you read them in order.

The introduction comes first.

Chapter 1 explains how to work with variables. We learn about number properties and the order of operations. These are basic concepts fundamental to solving algebra problems.

Chapter 2 defines expressions. This chapter covers the definition of an expression, parts of expressions, writing and simplifying expressions, and substitution.

Chapter 3 describes equations. Equations represent a scale balanced with numbers. You will learn parts of equations and how to simplify and solve them.

Chapter 4 introduces exponents. Exponents tell us how many times a number is multiplied by itself. We'll examine exponents of positive and negative numbers, operations with exponents, and negative exponents.

Chapter 5 delves into the concept of roots. We will study square roots, cube roots, and operations with roots, which are the inverse (or opposite) of exponents.

Chapters 6 and 7 address polynomials and polynomial equations. Here we'll learn about parts and types of polynomials, how to simplify polynomials, and how to perform operations with them. We also learn about polynomial equations and how to solve them.

Chapter 8 deals with absolute value. We'll review the concept of absolute value, which shows how far a number is from zero. We'll learn about how this applies to variables so we can solve absolute value equations.

Chapter 9 explains inequalities. Inequalities are values that are not equal—one value may be less than or greater than the other. This chapter covers operations with inequalities and how to solve them.

Chapter 10 focuses on systems of equations. We'll learn how to solve systems of equations in two ways: by substitution and by combining.

Chapter 11 concludes with word problems. Word problem in algebra are similar to story problems in basic math. In this chapter, some common types of word problems are explained. We'll also look at how to solve them.

Practice makes perfect!

As you read each chapter, you'll find practice exercises along the way. Complete the exercises to practice what you've learned. The more you use your skills, the better they will "stick" in your mind.

To learn more

Ready for more practice? After you've finished this book, visit the egghead website at http://www.petersonspublishing.com/egghead.aspx. Click the egghead link for even more practice exercises. This book will get you off to a great start. The website can give you that extra algebra boost!

Peterson's books

Along with producing the egghead's Guides, Peterson's publishes many types of books. These can help you prepare for tests, choose a college, and plan your career. They can even help you obtain financial aid. Look for Peterson's books at your school guidance office, local library or bookstore, or at www.petersonsbooks.com. Peterson's books are now available as ebooks, too!

We welcome any comments or suggestions you may have about this book. Your feedback will help us make educational dreams possible for you—and others like you.

Now that you know what's ahead, let's get started!

Introduction

What's the first thing that comes to your mind when you think of algebra?

In the past, you may have seen story problems with Train A traveling at 60 mph and Train B at 40 mph, with a question that asks whether Train A or Train B will arrive first. Solving these types of problems may make algebra seem tedious and downright irrelevant. We're not going to kid you— there are some problems like that in this book. They have to be included to prepare you for what you'll face on tests and in school.

Algebra is a standard course in every high school math curriculum. You have to learn it to move on to more advanced courses like geometry and trig. Most standardized tests include algebra problems too, so knowing this subject is a must for scoring well on the GED®, ACT®, or SAT*. Surprisingly, many jobs require some algebra skills, such as video game developer, animator, investment banker, business owner, and more.

But learning all of this isn't just for the purpose of getting a good job or better grades. In reality, we use this type of math reasoning often. Many everyday situations compel us to find numbers when we are missing some information. You might need to figure out, for example, if the quarter tank of gas in your car will be enough to reach your destination 20 miles away. Knowing how to solve an algebra equation can really come in handy for this! We use algebra every time we figure out how much to tip a restaurant server, or how much we'll save by purchasing an item on sale. We use it to decide how long we'll need to save to buy a new car, afford a nicer apartment, or take a special vacation. Often, we solve these types of problems in our minds without even realizing we're using algebra.

This book starts with the basics and progresses to more advanced material. If you're new to the subject, it will teach you algebra's main concepts. If you've studied algebra before but are a little rusty, it will provide a good review. Even if you don't plan to work as an engineer or a computer analyst someday, what you learn here can help you be a better problem solver in many areas of life. You may not become an algebra whiz, but hopefully you'll feel more confident tackling these types of questions. And you may even decide to major in math!

Let's begin! Meet egghead!

Chapter 1

Working with Variables

Hi! I'm egghead. In this chapter, we'll review the following concepts:

Commutative property
Associative property
Distributive property
Additive identity property
Multiplicative identity property
Additive inverse property
Multiplicative inverse property
Zero property
Order of operations

Number properties

In math, we use algebra to help us find missing numbers. These missing numbers are called **unknowns**.

Many people have a hard time with algebra because letters appear in the math problems. These letters are called **variables**.

An unknown is a number that we can find, if we have the right information.

Variable is a fancy word for "missing number."

We use variables in algebra to help us find missing numbers, or unknowns. That's really what algebra is all about. Here's how it works.

Say there is a number, but we don't know what the number is. We use a letter to represent the number—let's call it *x*.

$$7x = 14$$
$$x = unknown$$

The only thing we know is that if we multiply *x* by 7, the result is 14. We can use that information to find the value of *x*.

Before we start to find missing values, there are a few rules you'll need to know. These rules have to do with how we work with numbers. They are called **number properties**. We'll review eight **number properties** in this chapter.

1. Commutative property

The **commutative property** tells us that we can switch around the order of numbers when we add or multiply. The answer comes out the same.

Examples

If we add two numbers, 5 + 3, we can also add them this way: 3 + 5.

Here's how you might see the commutative property written:

3 + 5 = 5 + 3

This property works with multiplication, too:

3 x 5 = 5 x 3

Both answers are 15.

If we don't know the numbers we're working with, we can use variables:

$$a + b = b + a$$

Variables can be multiplied, too:

$$a \times b = b \times a$$

We might even add a number and a variable:

$$7 + y = y + 7$$

The commutative property doesn't work with division or subtraction. But we can use it with addition or multiplication any time. Try these practice questions to reinforce your skills.

Practice Questions—Commutative property

Directions: Show the commutative property for each operation. You will find the Practice Question Solutions on page 21.

1. $3 + 4 =$

2. $7 \times 9 =$

3. $479 + 25 =$

4. $66 \times 41 =$

5. $21 \times 974 =$

6. $a + b =$

7. $m + r =$

8. $k \times w =$

9. $l + q =$

10. $j \times t =$

2. Associative property

The **associative property** concerns the use of parentheses. Parentheses can be moved, but this doesn't change the answer.

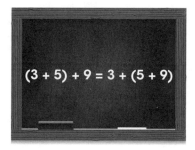

$$(3 + 5) + 9 = 3 + (5 + 9)$$

If you were to add these numbers, you would get the same answer on both sides:

$$(3 + 5) + 9 = 3 + (5 + 9)$$
$$(8) + 9 = 3 + (14)$$
$$17 = 17$$

In this example, we're changing the grouping of the numbers. On the left side, 3 and 5 are added in parentheses. On the right side, 5 and 9 are added. The grouping changes, but the answer stays the same.

Examples

Here's an example of the associative property using multiplication:

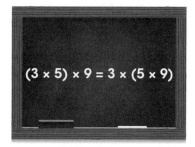

$$(3 \times 5) \times 9 = 3 \times (5 \times 9)$$

Both answers are 135:

$$(3 \times 5) \times 9 = 3 \times (5 \times 9)$$
$$(15) \times 9 = 3 \times (45)$$
$$135 = 135$$

Here's what it looks like with variables:

$$(a + b) + c = a + (b + c)$$

And here are some variables multiplied:

$$(a \times b) \times c = a \times (b \times c)$$

We can use the associative property with numbers and variables, too:

$$(7 + y) + 4 = 7 + (y + 4)$$

If you change the grouping, the answer stays the same. Again, this only works for addition and multiplication, not subtraction or division.

Try some associative property questions.

Practice Questions—

Associative property

Directions: Show the associative property for each set of operations. You will find the Practice Question Solutions on page 21.

11. $(1 + 2) + 5 =$

12. $(22 \times 5) \times 10 =$

13. $(15 + 6) + 3 =$

14. $(6 \times 5) \times 2 =$

15. $(4 + 68) + 11 =$

16. $(c + d) + e =$

17. $(a \times y) \times m =$

18. $(q + g) + s =$

19. $(b \times n) \times w =$

20. $(k \times z) \times p =$

egghead's Guide to Algebra

3. Distributive property

The **distributive property** describes how we multiply a number by two numbers being added. Say you have the following operation:

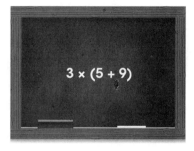

$$3 \times (5 + 9)$$

In this case, we would multiply 3 × 5 and 3 × 9. Then, add those products together:

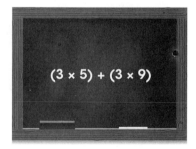

$$(3 \times 5) + (3 \times 9)$$

This is called **distributing the 3**.

Examples

We can use the distributive property with variables:

$$a \times (b + c) = (a \times b) + (a \times c)$$

This might also be written without the multiplication sign:

$$a(b + c) = (a \times b) + (a \times c)$$
$$= (ab) + (ac)$$
$$= ab + ac$$

When the variables are placed right next to each other, this means they are multiplied.

$$(a \times b) = ab$$

Here's an example of the distributive property with numbers and variables:

$$7 \times (y + 4) = (7 \times y) + (7 \times 4)$$

In this case, we distribute the 7 over the y and the 4.

Practice Questions—Distributive property

Directions: Show the distributive property for each set of operations. You will find the Practice Question Solutions on page 21.

21. $6(3 + 7) =$

22. $9(1 + 4) =$

23. $2(5 + 6) =$

24. $8(1 + 2) =$

25. $4(25 + 30) =$

26. $c(d + e) =$

27. $u(r + z) =$

28. $s(n + o) =$

29. $q(a + x) =$

30. $g(b + k) =$

4. Additive identity

The commutative, associative, and distributive properties are the main properties you'll need to know to solve problems using algebra. But there are a few other properties that will come in handy as well. These are the identity, inverse, and zero properties.

The identity properties tell us about what numbers you can add or multiply to keep the original number the same.

Let's start with an original number of 10. What number, when added to 10, gives us 10?

$$10 + \underline{\hspace{2cm}} = 10$$

The answer is 0.

$$10 + 0 = 10$$

The **additive identity** property tells us that when you add zero to a number, the value of the number doesn't change. This applies to variables as well:

$$a + 0 = a$$

The number stays the same, so it has the same identity.

This one is pretty straightforward, so we'll skip straight to practice.

Practice Questions—Additive identity

Directions: Fill in the blanks to show the additive identity property for each question below. You will find the Practice Question Solutions on page 21.

31. $14 + 0 = \underline{\hspace{2cm}}$

32. $\underline{\hspace{2cm}} + 0 = 10,000$

33. $75 + \underline{\hspace{2cm}} = 75$

34. $\underline{\hspace{2cm}} + 0 = 126$

35. $295 + 0 = \underline{\hspace{2cm}}$

36. $x + 0 = \underline{\hspace{2cm}}$

37. $a + \underline{\hspace{2cm}} = a$

38. $0 + w = \underline{\hspace{2cm}}$

39. $\underline{\hspace{2cm}} + v = v$

40. $g + 0 = \underline{\hspace{2cm}}$

5. Multiplicative identity

Multiplicative identity is similar to additive identity. What number, when multiplied by 10, gives us 10?

$$10 \times \underline{\hspace{1cm}} = 10$$

The answer is 1.

$$10 \times 1 = 10$$

The **multiplicative identity** property tells us that when you multiply a number by 1, the value of the number stays the same. Here's an example using a variable:

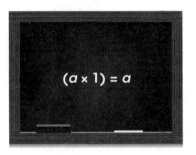

$$(a \times 1) = a$$

Practice Questions—Multiplicative identity

Directions: Fill in the blanks to show the multiplicative identity property for each question below. You will find the Practice Question Solutions on page 22.

41. $6 \times 1 =$ _____

42. _____ $\times 1 = 58{,}691$

43. $14 \times$ _____ $= 14$

44. _____ $\times 1 = 260$

45. $458 \times$ _____ $= 458$

46. $c \times 1 =$ _____

47. $1 \times$ _____ $= s$

48. _____ $\times 1 = q$

49. $w \times 1 =$ _____

50. $y \times$ _____ $= y$

6. Additive inverse

The **additive inverse** property tells us what to add to a number if we want to get a result of 0.

$$10 + (-10) = 0$$

To get zero, we add the equivalent negative number.

Examples

The additive inverse works with numbers and variables:

$$15 + (-15) = 0$$
$$a + (-a) = 0$$
$$3 + (-3) = 0$$
$$p + (-p) = 0$$

It doesn't matter what number or variable you start with. If you add its opposite, the result is zero.

Practice Questions—Additive inverse

Directions: Fill in the blanks to show the additive inverse property for each question below. You will find the Practice Question Solutions on page 22.

51. $17 + (-17) =$ _____

52. _____ $+ (-356) = 0$

53. $2 +$ _____ $= 0$

54. _____ $+ (-58) = 0$

55. $95 + (-95) =$ _____

56. $x + (-x) =$ _____

57. _____ $+ (-h) = 0$

58. $y +$ _____ $= 0$

59. _____ $+ (-s) = 0$

60. $r + (-r) =$ _____

7. Multiplicative inverse

The **multiplicative inverse** property is similar to the additive inverse property. It tells us what to multiply a number by to get a result of 1.

$$10 \times \frac{1}{10} = 1$$

To get a result of 1, we multiply the number by a fraction. The fraction is always 1 over the original number.

This fraction is called the **reciprocal** of the number.

Examples

Here are some examples of the multiplicative inverse property with numbers:

$$2 \times \frac{1}{2} = 1$$

$$4 \times \frac{1}{4} = 1$$

$$15 \times \frac{1}{15} = 1$$

Here are some examples using variables:

$$a \times \frac{1}{a} = 1$$

$$z \times \frac{1}{z} = 1$$

$$24r \times \frac{1}{24r} = 1$$

The product of the numbers is always 1.

Practice Questions—Multiplicative inverse

Directions: Fill in the blanks to show the multiplicative inverse property for each question below. You will find the Practice Question Solutions on page 22.

61. $4 \times$ _____ $= 1$

62. $89 \times$ _____ $= 1$

63. $51 \times$ _____ $= 1$

64. $7 \times$ _____ $= 1$

65. $21 \times$ _____ $= 1$

66. $y \times$ _____ $= 1$

67. $a \times$ _____ $= 1$

68. $f \times$ _____ $= 1$

69. $s \times$ _____ $= 1$

70. $w \times$ _____ $= 1$

8. Zero property

There is one more property to know about, called the zero property.

This property might be the easiest one to remember. It tells us that $2 \times 0 = 0$. We also know that $40,000 \times 0 = 0$. Even $1,000,000,000 \times 0$ is, of course, 0.

Any number multiplied by 0 equals 0.

Practice Questions—Zero property

Directions: Fill in the blanks to show the zero property for each question below. You will find the Practice Question Solutions on page 22.

71. $11 \times 0 = $ _____

72. $5 \times 0 = $ _____

73. $123 \times $ _____ $= 0$

74. $60,000 \times $ _____ $= 0$

75. $789 \times 0 = $ _____

76. $r \times 0 = $ _____

77. $f \times 0 = $ _____

78. $k \times $ _____ $= 0$

79. $p \times $ _____ $= 0$

80. $t \times 0 = $ _____

This practice should help you remember the properties. You'll have more opportunity to practice in the Chapter Review.

Order of operations

The operations we have worked with so far have been simple. When multiplying 5 × 3 × 9, you multiply from left to right, starting with the 5. But what if we come across a more complex operation?

How do we know which operations to perform first?

The order of operations turns out to be very important in math. There is a rule that tells us which operations to perform in what order. It is known as PEMDAS.

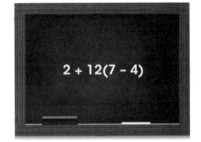

The P in PEMDAS stands for "parentheses." Operations in parentheses are performed first.

$$2 + 12(7 - 4) = 2 + 12(3)$$

Here we subtracted 7 − 4 to get 3. Operations involving exponents are performed second.

The E in PEMDAS stands for "exponents."

There are no exponents in this example, so we'll move to the next step.

The M stands for "multiplication," and the D stands for "division." Multiplication and division are performed next, in order from left to right.

$$2 + 12(7 - 4) = 2 + 12(3)$$
$$= 2 + 36$$

Here we multiplied 12 × 3. The product is 36.

The A in PEMDAS stands for "addition," and the S stands for "subtraction." Addition and subtraction are performed last, from left to right:

$$2 + 12(7 - 4) = 2 + 12(3)$$
$$= 2 + 36$$
$$= 38$$

The correct answer is 38.

egghead's Guide to Algebra

Notice what would have happened if we had not followed PEMDAS. If we had just performed every operation in order from left to right, we would have gotten an incorrect answer. Here's what the wrong solution would look like:

Step	Result
Add 2 + 12	2 + 12 = 14
Multiply by 7	14 × 7 = 98
Subtract 4	98 – 4 = 94
Wrong answer	94

If we had not followed PEMDAS, we might have found an answer of 94.

That's incorrect!

There is an easy way to remember PEMDAS. It's a saying we use as a memory device:

This phrase can help you remember: <u>P</u>arentheses, <u>E</u>xponents, <u>M</u>ultiplication, <u>D</u>ivision, <u>A</u>ddition, and <u>S</u>ubtraction.

Practice Questions—Order of operations

Directions: Show the correct order of operations to solve each question below. You will find the Practice Question Solutions on page 23.

81. $9 - 2 + 7 =$

82. $11 + (8 \times 2) \div 4 =$

83. $21 - 6(4 + 2) =$

84. $5 \times 4 + 5 \times 4 =$

85. $12 \div 3 + 16 =$

86. $20 + (6 \times 5) \div 5 =$

87. $8 \times 7 + 6 \times 5 =$

88. $60 \div 3 + 10 =$

89. $20 - 10(10 + 2) =$

90. $85 - (7 \times 2) =$

Chapter Review

Solutions can be found on page 24.

1. Show the commutative property for this operation.

   $95 + 4 =$

2. Show the commutative property for this operation.

   $s \times u =$

3. Show the commutative property for this operation.

   $c + f =$

4. Show the commutative property for this operation.

   $46 + 15 =$

5. Show the commutative property for this operation.

   $6 \times 9 =$

6. Show the associative property for this set of operations.

   $(15 + 8) + 75 =$

7. Show the associative property for this set of operations.

   $(p + n) + g =$

8. Show the associative property for this set of operations.

   $(2 \times 4) \times 150 =$

9. Show the associative property for this set of operations.

   $(r \times z) \times w =$

10. Show the associative property for this set of operations.

    $(h + t) + a =$

11. Show the distributive property for this set of operations.

    $12(4 + 6) =$

12. Show the distributive property for this set of operations.

    $d(k + e) =$

13. Show the distributive property for this set of operations.

$61(2 + 8) =$

14. Show the distributive property for this set of operations.

$r(j + a) =$

15. Show the distributive property for this set of operations.

$7(2 + 45) =$

16. Fill in the blank to show the additive identity property.

$105 + \underline{\hspace{1cm}} = 105$

17. Fill in the blank to show the multiplicative identity property.

$\underline{\hspace{1cm}} \times 1 = 29$

18. Fill in the blank to show the additive inverse property.

$91 + (-91) = \underline{\hspace{1cm}}$

19. Fill in the blank to show the multiplicative inverse property.

$a \times \underline{\hspace{1cm}} = 1$

20. Fill in the blank to show the zero property.

$b \times \underline{\hspace{1cm}} = 0$

21. Show the correct order of operations to solve the problem.

$5 + (9 \times 45) \div 15 =$

22. Show the correct order of operations to solve the problem.

$3 \times 2 + 2 \times 3 =$

23. Show the correct order of operations to solve the problem.

$185 - 10 + 95 =$

24. Show the correct order of operations to solve the problem.

$281 - 15(7 + 1) =$

25. Show the correct order of operations to solve the problem.

$18 - 11(5 + 9) =$

Practice Question Solutions

Commutative property

1. $3 + 4 = 4 + 3$

2. $7 \times 9 = 9 \times 7$

3. $479 + 25 = 25 + 479$

4. $66 \times 41 = 41 \times 66$

5. $21 \times 974 = 974 \times 21$

6. $a + b = b + a$

7. $m + r = r + m$

8. $k \times w = w \times k$

9. $l + q = q + l$

10. $j \times t = t \times j$

Associative property

11. $(1 + 2) + 5 = 1 + (2 + 5)$

12. $(22 \times 5) \times 10 = 22 \times (5 \times 10)$

13. $(15 + 6) + 3 = 15 + (6 + 3)$

14. $(6 \times 5) \times 2 = 6 \times (5 \times 2)$

15. $(4 + 68) + 11 = 4 + (68 + 11)$

16. $(c + d) + e = c + (d + e)$

17. $(a \times y) \times m = a \times (y \times m)$

18. $(q + g) + s = q + (g + s)$

19. $(b \times n) \times w = b \times (n \times w)$

20. $(k \times z) \times p = k \times (z \times p)$

Distributive property

21. $6(3 + 7) = (6 \times 3) + (6 \times 7)$

22. $9(1 + 4) = (9 \times 1) + (9 \times 4)$

23. $2(5 + 6) = (2 \times 5) + (2 \times 6)$

24. $8(1 + 2) = (8 \times 1) + (8 \times 2)$

25. $4(25 + 30) = (4 \times 25) + (4 \times 30)$

26. $c(d + e) = (c \times d) + (c \times e)$

27. $u(r + z) = (u \times r) + (u \times z)$

28. $s(n + o) = (s \times n) + (s \times o)$

29. $q(a + x) = (q \times a) + (q \times x)$

30. $g(b + k) = (g \times b) + (g \times k)$

Additive identity

31. $14 + 0 = 14$

32. $10,000 + 0 = 10,000$

33. $75 + 0 = 75$

34. $126 + 0 = 126$

35. $295 + 0 = 295$

36. $x + 0 = x$

37. $a + 0 = a$

38. $0 + w = w$

39. $0 + v = v$

40. $g + 0 = g$

Multiplicative identity

41. $6 \times 1 = 6$

42. $58,691 \times 1 = 58,691$

43. $14 \times 1 = 14$

44. $260 \times 1 = 260$

45. $458 \times 1 = 458$

46. $c \times 1 = c$

47. $1 \times s = s$

48. $q \times 1 = q$

49. $w \times 1 = w$

50. $y \times 1 = y$

Additive inverse

51. $17 + (-17) = 0$

52. $356 + (-356) = 0$

53. $2 + (-2) = 0$

54. $58 + (-58) = 0$

55. $95 + (-95) = 0$

56. $x + (-x) = 0$

57. $h + (-h) = 0$

58. $y + (-y) = 0$

59. $s + (-s) = 0$

60. $r + (-r) = 0$

Multiplicative inverse

61. $4 \times \frac{1}{4} = 1$

62. $89 \times \frac{1}{89} = 1$

63. $51 \times \frac{1}{51} = 1$

64. $7 \times \frac{1}{7} = 1$

65. $21 \times \frac{1}{21} = 1$

66. $y \times \frac{1}{y} = 1$

67. $a \times \frac{1}{a} = 1$

68. $f \times \frac{1}{f} = 1$

69. $s \times \frac{1}{s} = 1$

70. $w \times \frac{1}{w} = 1$

Zero property

71. $11 \times 0 = 0$

72. $5 \times 0 = 0$

73. $123 \times 0 = 0$

74. $60,000 \times 0 = 0$

75. $789 \times 0 = 0$

76. $r \times 0 = 0$

77. $f \times 0 = 0$

78. $k \times 0 = 0$

79. $p \times 0 = 0$

80. $t \times 0 = 0$

Order of operations

81. The correct answer is 14.

$$9 - 2 + 7 = 7 + 7$$
$$= 14$$

82. The correct answer is 15.

$$11 + (8 \times 2) \div 4 = 11 + 16 \div 4$$
$$= 11 + 4$$
$$= 15$$

83. The correct answer is -15.

$$21 - 6(4 + 2) = 21 - 6(6)$$
$$= 21 - 36$$
$$= -15$$

84. The correct answer is 40.

$$5 \times 4 + 5 \times 4 = 20 + 20$$
$$= 40$$

85. The correct answer is 20.

$$12 \div 3 + 16 = 4 + 16$$
$$= 20$$

86. The correct answer is 26.

$$20 + (6 \times 5) \div 5 = 20 + 30 \div 5$$
$$= 20 + 6$$
$$= 26$$

87. The correct answer is 86.

$$8 \times 7 + 6 \times 5 = 56 + 30$$
$$= 86$$

88. The correct answer is 30.

$$60 \div 3 + 10 = 20 + 10$$
$$= 30$$

89. The correct answer is -100.

$$20 - 10(10 + 2) = 20 - 10(12)$$
$$= 20 - 120$$
$$= -100$$

90. The correct answer is 71.

$$85 - (7 \times 2) = 85 - 14$$
$$= 71$$

**Chapter
Review
Solutions**

1. $95 + 4 = 4 + 95$

2. $s \times u = u \times s$

3. $c + f = f + c$

4. $46 + 15 = 15 + 46$

5. $6 \times 9 = 9 \times 6$

6. $(15 + 8) + 75 = 15 + (8 + 75)$

7. $(p + n) + g = p + (n + g)$

8. $(2 \times 4) \times 150 = 2 \times (4 \times 150)$

9. $(r \times z) \times w = r \times (z \times w)$

10. $(h + f) + a = h + (t + a)$

11. $12(4 + 6) = (12 \times 4) + (12 \times 6)$

12. $d(k + e) = (d \times k) + (d \times e)$

13. $61(2 + 8) = (61 \times 2) + (61 \times 8)$

14. $r(j + a) = (r \times j) + (r \times a)$

15. $7(2 + 45) = (7 \times 2) + (7 \times 45)$

16. $105 + 0 = 105$

17. $29 \times 1 = 29$

18. $91 + (-91) = 0$

19. $a \times \dfrac{1}{a} = 1$

20. $b \times 0 = 0$

21. The correct answer is 32.
$$\begin{aligned} 5 + (9 \times 45) \div 15 &= 5 + 405 \div 15 \\ &= 5 + 27 \\ &= 32 \end{aligned}$$

22. The correct answer is 12.
$$\begin{aligned} 3 \times 2 + 2 \times 3 &= 6 + 6 \\ &= 12 \end{aligned}$$

23. The correct answer is 270.
$$\begin{aligned} 185 - 10 + 95 &= 175 + 95 \\ &= 270 \end{aligned}$$

24. The correct answer is 161.
$$\begin{aligned} 281 - 15(7 + 1) &= 281 - 15(8) \\ &= 281 - 120 \\ &= 161 \end{aligned}$$

25. The correct answer is -136.
$$\begin{aligned} 18 - 11(5 + 9) &= 18 - 11(14) \\ &= 18 - 154 \\ &= -136 \end{aligned}$$

Chapter 2

Expressions

In this chapter, we'll review the following concepts:

What is an expression?
Parts of expressions
Writing expressions
Simplifying expressions
Substitution

What is an expression?

An algebra **expression** is a math statement containing numbers, variables, and operations.

Here are a few examples:

$$2x + 5$$

$$17 - 24a$$

$$3(2n + 9) - 74$$

We work with expressions throughout algebra, so let's learn more about them.

It does not contain an equal sign.

Parts of expressions

Expressions are made up of four parts. First are **variables**, which are letters that represent unknown numbers. In the expression shown, x is a variable.

$$2x + 5$$

Second are the **constants**. Constants are fixed numbers with known values. In the expression $2x + 5$, the number 5 is a constant.

The third parts of expressions are called **coefficients**. Coefficients are numbers that are multiplied by variables. In the expression $2x + 5$, the number 2 is a coefficient.

It is multiplied by the variable x.

Expressions also contain **mathematical symbols** that show the operation being performed. In the expression $2x + 5$, the + sign is the mathematical symbol. Other symbols might show subtraction, multiplication, or division.

Writing expressions

To create expressions, we sometimes need to translate written ideas to math. For example, we might be asked to find the sum of 12 and an unknown number, y.

That would be written as $12 + y$.

The word "sum" tells us the numbers are being added.

We might also be asked something more complicated:

Find the difference between two times a and four times b.

This would be written as $2a - 4b$. The phrase "difference between" tells us we're subtracting, and the word "times" tells us to multiply. Here's a list of common operation words and how to translate them into math.

Word or phrase	What it means	Symbol to use
difference	Subtraction	–
equals	Two values are the same	=
increased by	Addition	+
is	The result of an operation	=
less than	Subtraction	–
minus	Subtraction	–
more than	Addition	+
plus	Addition	+

Word or phrase	What it means	Symbol to use
product	Multiplication	×
quotient	Division	÷
reduced by	Subtraction	−
sum	Addition	+
times	Multiplication	×

You may come across different math words than the ones shown here, but learning these is a good place to start.

Practice Questions—Writing expressions

Directions: Write the algebra expressions for the phrases below. You will find the Practice Question Solutions on page 39.

1. The sum of x and y

2. The product of a and b

3. The quotient of $5c$ and two

4. Seven less than r

5. Five more than z

6. The difference between $4b$ and c

7. The product of x, y, and z

8. The sum of nine and k

9. Seven increased by x

10. Twelve divided by n

Simplifying expressions

To work with expressions, we start by simplifying them. **Simplifying** means putting an expression into its simplest form.

To simplify expressions, we take two steps. First, we use the distributive property.

Then, we combine like terms. Here's how it works.

This eliminates the parentheses.

Expressions with one variable

If you can distribute any numbers, do that first.

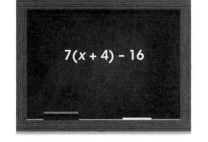

$$7(x + 4) - 16$$

In the example above, we would first distribute the 7, as shown:

$$7(x + 4) - 16$$

$$(7 \times x) + (7 \times 4) - 16$$

$$7x + 28 - 16$$

Next, combine like terms. **Like terms** are the parts of an expression that are similar. They might have the same variable, or they might both be constants. In this case, there is only one term with a variable: $7x$. The numbers 28 and 16 are like terms, however, so we combine them:

$$7x + 28 - 16$$

$$7x + 12$$

This is the expression in its simplest form.

If an expression has two terms with variables, you can combine them, too.

The following expression has two terms with variables:

$$5a + 12 + 6a$$

To combine like terms with variables, add the coefficients.

Both terms have the same variable, a, so they're like terms. To combine them, we add $5 + 6$:

$$5a + 6a + 12$$

$$11a + 12$$

Try it now on your own.

Practice Questions—Expressions with one variable

Directions: Simplify the expressions below. You will find the Practice Question Solutions on page 39.

11. $2a + 4a$

12. $7y + 19y$

13. $3x + 40x + 65$

14. $12 - 3r + 7$

15. $12a \times 3$

16. $7r - 10r$

17. $15b - 2 + 5b$

18. $3x + 5x + 4$

19. $9 + 3a - 2$

20. $3x + 5 + 6x + 2$

Expressions with two variables

We can simplify expressions with two or more variables also. It helps to move like terms next to each other before combining.

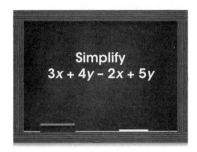

Step 1: Use the distribution property.

In this expression, there's nothing to distribute, so we'll move to the next step.

Step 2: Combine like terms.

This expression contains terms with two variables, x and y. Move the like terms next to each other:

$$3x + 4y - 2x + 5y$$

$$3x - 2x + 4y + 5y$$

Now, add or subtract:

$$3x - 2x + 4y + 5y$$

$$x + 4y + 5y$$

$$x + 9y$$

Only terms with the same variables can be combined.

This is as simple as this expression will get. You cannot combine x with $9y$, because these terms have different variables.

Practice Questions—Expressions with two variables

Directions: Simplify the expressions below. You will find the Practice Question Solutions on page 40.

21. $6d + 7e + 3d + 2e$

22. $14(v + 2z)$

23. $76 - 3p + 2q + 4p$

24. $2(3a - 4b)$

25. $15r + 9 + 6s + 2 - 3r$

26. $x + 2y + 5 + 2x - 3y$

27. $x - y + 2x + 2y - 11$

28. $2(x - y) + 2x$

29. $2r + 4(2r + 3s) - 2s$

30. $4(a + b) + 7(2a - 3b)$

Substitution

Once we simplify an expression, we can find its value using substitution. **Substitution** involves replacing a variable with a given number.

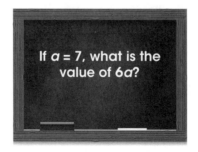

If $a = 7$, what is the value of $6a$?

To answer this question, substitute 7 for *a* in the expression. Then multiply:

$$6a = 6 \times a$$
$$= 6 \times (7)$$
$$= 42$$

The value of $6a$ is 42.

Substitution for one variable

The process of substitution is also known as **evaluation**. When we evaluate an expression, we're just plugging in a number for a variable.

Here are more examples of how it's done.

What is the value of 9*x* + 4, if *x* = 2?

Substitute 2 for *x* in the expression:

$$9x + 4 = 9(2) + 4$$
$$= 18 + 4$$
$$= 22$$

If *x* equals 2, then $9x + 4$ equals 22.

What is the value of 3*c* + 2*c*, if *c* = 9?

Here, we plug in 9 for the variable *c* both times:

$$3c + 2c = 3(9) + 2(9)$$
$$= 27 + 18$$
$$= 45$$

The value of the expression is 45.

Practice Questions—Substitution for one variable

Directions: Find the value of the expressions below using substitution. You will find the Practice Question Solutions on page 40.

31. If $x = 2$, what is $2x + 3$?

32. Find the value of $4k + 12$. Substitute 6 for k.

33. Find the value of $5m + 4m$ if $m = 3$.

34. If $r = 7$, what is the value of $3r + 5$?

35. Find the value of $12z \div 6$, if $z = 4$.

36. Given that $s = 2$, what is $14s - 2$?

37. If $n = 3$, find the value of $5n \times 7$.

38. Find the value of $x + 14x + 2$, given that $x = 5$.

39. What is the value of $2k - 11$, if $k = 6$?

40. If $r = 9$, find the value of $\dfrac{15r + 6r}{3}$.

Substitution for two variables

We can evaluate expressions with two or more variables as well. We just need to know which number to substitute for which variable.

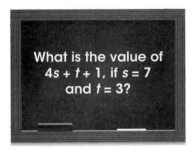

Plug in 7 for s and 3 for t in the expression. Then do the math:

$$4s + t + 1 = 4(7) + (3) + 1$$
$$= 28 + 3 + 1$$
$$= 31 + 1$$
$$= 32$$

The value of the expression is 32.

If $r = 5$ and $p = 40$, find the value of $90 - 2(r + p)$.

Substitute 5 for r and 40 for p in the expression, as shown:

$$90 - 2(r + p) = 90 - 2(5 + 40)$$
$$= 90 - 2(45)$$
$$= 90 - 90$$
$$= 0$$

The correct answer is 0.

Practice Questions—Substitution for two variables

Directions: Determine the value of the expressions below using substitution. You will find the Practice Question Solutions on page 41.

41. Find the value of $3x + 2y$. Substitute 8 for x and 7 for y.

42. What is the value of $10m + 10n$ if $m = 4$ and $n = 5$?

43. If $a = 1$ and $b = 3$, what is the value of $a + 2b$?

44. Given that $k = 4$ and $m = 3$, find the value of $5(2k + 7m)$.

45. What is the value of $(15x + 21y) \div 3$, if 2 is substituted for x and 7 is substituted for y?

46. If $a = 5$ and $b = 0$, what is the value of $3a + 2(a + 4b)$?

47. Find the value of $10(y + 4z)$, given that $y = 5$ and $z = 3$.

48. What is the value of $(14m + 20n) \div 2$ if the value of m is 6, and the value of n is 7?

49. Given that $x = 8$ and $y = 2$, determine the value of $7x - 8y$.

50. Find the value of $5(3r - 4s)$, if $r = 9$ and $s = 3$.

Chapter Review

Solutions can be found on page 42.

1. Write the algebra expression for the phrase below.

   The sum of m and n

2. Write the algebra expression for the phrase below.

   The product of $6r$ and 7

3. Write the algebra expression for the phrase below.

   The difference between $3x$ and $4y$

4. Write the algebra expression for the phrase below.

   The product of k and m

5. Write the algebra expression for the phrase below.

   The quotient of $8s$ and 3

6. Simplify the expression shown.

   $7x + 9x$

7. Simplify the expression shown.

   $2r + 16 - r$

8. Simplify the expression shown.

   $12m \times 3$

9. Simplify the expression shown.

   $3 + 6n - 4$

10. Simplify the expression shown.

    $3z + 5z - 4$

11. Simplify the expression shown. What number property is used to simplify the expression?

    $11(m + 2)$

12. Simplify the expression shown. Combine like terms for both variables.

    $56 - 4y + 2z + 6y$

13. Simplify the expression shown. Combine like terms for both variables.

    $15r + 6s + 2r - 3s$

14. Simplify the expression shown. Combine like terms for both variables.

    $3m - 4n + 6m + 3n + 6$

15. Simplify the expression shown. Combine like terms for both variables.

$4(k + m) + 8(2k - 3m)$

16. Find the value of $7z \div 2$, if $z = 4$.

17. If $m = 3$, what is the value of $9m - 2$?

18. If $k = 3$, find the value of $4k - 11$.

19. Find the value of $3x + 17x + 5$, given that $x = 4$.

20. What is the value of $2z \times 4$, if $z = 18$?

21. What is the value of $(5x + 2y) \div 3$, if 3 is substituted for x and 6 is substituted for y?

22. If $m = 7$ and $n = 2$, what is the value of $5m + 3(m + 4n)$?

23. Find the value of $8(r + 6s)$, if $r = 4$ and $s = 6$.

24. What is the value of $(12y + 16z) \div 4$ if the value of y is 3, and the value of z is 5?

25. Given that $a = 7$ and $b = 3$, determine the value of $5a - 9b$.

Practice Question Solutions

Writing expressions

1. $x + y$

2. $a \times b$

3. $5c \div 2$

4. $r - 7$

5. $z + 5$

6. $4b - c$

7. xyz

8. $9 + k$

9. $7 + x$

10. $12 \div n$

Expressions with one variable

11. The correct answer is $6a$.

   $2a + 4a$

   $6a$

12. The correct answer is $26y$.

   $7y + 19y$

   $26y$

13. The correct answer is $43x + 65$.

   $3x + 40x + 65$

   $43x + 65$

14. The correct answer is $19 - 3r$.

   $12 - 3r + 7$

   $12 + 7 - 3r$

   $19 - 3r$

15. The correct answer is $36a$.

   $12a \times 3$

   $36a$

16. The correct answer is $-3r$.

   $7r - 10r$

   $-3r$

17. The correct answer is $20b - 2$.

   $15b - 2 + 5b$

   $15b + 5b - 2$

   $20b - 2$

18. The correct answer is $8x + 4$.

   $3x + 5x + 4$

   $8x + 4$

19. The correct answer is $7 + 3a$.

   $9 + 3a - 2$

   $9 - 2 + 3a$

   $7 + 3a$

20. The correct answer is $9x + 7$.

   $3x + 5 + 6x + 2$

   $3x + 6x + 5 + 2$

   $9x + 7$

Expressions with two variables

21. The correct answer is $9d + 9e$.

$6d + 7e + 3d + 2e$

$6d + 3d + 7e + 2e$

$9d + 9e$

22. The correct answer is $14v + 28z$.

$14(v + 2z)$

$(14 \times v) + (14 \times 2z)$

$(14v) + (14 \times 2z)$

$14v + 28z$

23. The correct answer is $76 + p + 2q$.

$76 - 3p + 2q + 4p$

$76 - 3p + 4p + 2q$

$76 + p + 2q$

24. The correct answer is $6a - 8b$.

$2(3a - 4b)$

$(2 \times 3a) - (2 \times 4b)$

$(6a) - (2 \times 4b)$

$6a - 8b$

25. The correct answer is $12r + 11 + 6s$.

$15r + 9 + 6s + 2 - 3r$

$15r - 3r + 9 + 2 + 6s$

$12r + 9 + 2 + 6s$

$12r + 11 + 6s$

26. The correct answer is $3x - y + 5$.

$x + 2y + 5 + 2x - 3y$

$x + 2x + 2y - 3y + 5$

$3x + 2y - 3y + 5$

$3x - y + 5$

27. The correct answer is $3x + y - 11$.

$x - y + 2x + 2y - 11$

$x + 2x - y + 2y - 11$

$3x - y + 2y - 11$

$3x + y - 11$

28. The correct answer is $4x - 2y$.

$2(x - y) + 2x$

$(2 \times x) - (2 \times y) + 2x$

$(2x) - (2y) + 2x$

$2x + 2x - 2y$

$4x - 2y$

29. The correct answer is $10r + 10s$.

$2r + 4(2r + 3s) - 2s$

$2r + (4 \times 2r) + (4 \times 3s) - 2s$

$2r + (8r) + (12s) - 2s$

$10r + 10s$

30. The correct answer is $18a - 17b$.

$4(a + b) + 7(2a - 3b)$

$4(a) + 4(b) + (7 \times 2a) - (7 \times 3b)$

$4a + 4b + (14a) - (21b)$

$4a + 14a + 4b - 21b$

$18a - 17b$

Substitution for one variable

31. The correct answer is 7.

$$2x + 3 = 2(2) + 3$$
$$= 4 + 3$$
$$= 7$$

32. The correct answer is 36.

$$4k + 12 = 4(6) + 12$$
$$= 24 + 12$$
$$= 36$$

33. The correct answer is 27.

$$5m + 4m = 5(3) + 4(3)$$
$$= 15 + 12$$
$$= 27$$

34. The correct answer is 26.

$$3r + 5 = 3(7) + 5$$
$$= 21 + 5$$
$$= 26$$

35. The correct answer is 8.

$$12z \div 6 = 12(4) \div 6$$
$$= 48 \div 6$$
$$= 8$$

36. The correct answer is 26.

$$14s - 2 = 14(2) - 2$$
$$= 28 - 2$$
$$= 26$$

37. The correct answer is 105.

$$5n \times 7 = 5(3) \times 7$$
$$= 15 \times 7$$
$$= 105$$

38. The correct answer is 77.

$$x + 14x + 2 = (5) + 14(5) + 2$$
$$= 5 + 70 + 2$$
$$= 75 + 2$$
$$= 77$$

39. The correct answer is 1.

$$2k - 11 = 2(6) - 11$$
$$= 12 - 11$$
$$= 1$$

40. The correct answer is 63.

$$\frac{15r + 6r}{3} = \frac{15(9) + 6(9)}{3}$$
$$= \frac{135 + 54}{3}$$
$$= \frac{189}{3}$$
$$= 63$$

Substitution for two variables

41. The correct answer is 38.

$$3x + 2y = 3(8) + 2(7)$$
$$= 24 + 14$$
$$= 38$$

42. The correct answer is 90.

$$10m + 10n = 10(4) + 10(5)$$
$$= 40 + 50$$
$$= 90$$

43. The correct answer is 7.

$$a + 2b = (1) + 2(3)$$
$$= 1 + 6$$
$$= 7$$

44. The correct answer is 145.

$$5(2k + 7m) = 5[2(4) + 7(3)]$$
$$= 5(8 + 21)$$
$$= 5(29)$$
$$= 145$$

45. The correct answer is 59.

$$(15x + 21y) \div 3 = [15(2) + 21(7)] \div 3$$
$$= (30 + 147) \div 3$$
$$= 177 \div 3$$
$$= 59$$

46. The correct answer is 25.

$$3a + 2(a + 4b) = 3(5) + 2[(5) + 4(0)]$$
$$= 15 + 2(5 + 0)$$
$$= 15 + 2(5)$$
$$= 15 + 10$$
$$= 25$$

47. The correct answer is 170.

$$10(y + 4z) = 10[(5) + 4(3)]$$
$$= 10(5 + 12)$$
$$= 10(17)$$
$$= 170$$

48. The correct answer 112.

$$(14m + 20n) \div 2 = [14(6) + 20(7)] \div 2$$
$$= (84 + 140) \div 2$$
$$= 224 \div 2$$
$$= 112$$

49. The correct answer is 40.

$$7x - 8y = 7(8) - 8(2)$$
$$= 56 - 16$$
$$= 40$$

50. The correct answer is 75.

$$5(3r - 4s) = 5[3(9) - 4(3)]$$
$$= 5(27 - 12)$$
$$= 5(15)$$
$$= 75$$

Chapter Review Solutions

1. $m + n$

2. $6r \times 7$

3. $3x - 4y$

4. $k \times m$

5. $8s \div 3$

6. The correct answer is $16x$.

$7x + 9x$

$16x$

7. The correct answer is $r + 16$.

$2r + 16 - r$

$2r - r + 16$

$r + 16$

8. The correct answer is $36m$.

$12m \times 3$

$36m$

9. The correct answer is $6n - 1$.

$3 + 6n - 4$

$3 - 4 + 6n$

$-1 + 6n$

$6n - 1$

Chapter 2: Expressions

10. The correct answer is $8z - 4$.

$3z + 5z - 4$

$8z - 4$

11. The correct answer is $11m + 22$. The distributive property is used to simplify the expression:

$11(m + 2)$

$(11 \times m) + (11 \times 2)$

$11m + (11 \times 2)$

$11m + 22$

12. The correct answer is $56 + 2y + 2z$.

$56 - 4y + 2z + 6y$

$56 - 4y + 6y + 2z$

$56 + 2y + 2z$

13. The correct answer is $17r + 3s$.

$15r + 6s + 2r - 3s$

$15r + 2r + 6s - 3s$

$17r + 6s - 3s$

$17r + 3s$

14. The correct answer is $9m - n + 6$.

$3m - 4n + 6m + 3n + 6$

$3m + 6m - 4n + 3n + 6$

$9m - n + 6$

15. The correct answer is $20k - 20m$.

$4(k + m) + 8(2k - 3m)$

$(4 \times k) + (4 \times m) + (8 \times 2k) - (8 \times 3m)$

$(4k) + (4m) + (16k) - (24m)$

$4k + 16k + 4m - 24m$

$20k + 4m - 24m$

$20k - 20m$

16. The correct answer is 14.

$$7z \div 2 = 7(4) \div 2$$
$$= 28 \div 2$$
$$= 14$$

17. The correct answer is 25.

$$9m - 2 = 9(3) - 2$$
$$= 27 - 2$$
$$= 25$$

18. The correct answer is 1.

$$4k - 11 = 4(3) - 11$$
$$= 12 - 11$$
$$= 1$$

19. The correct answer is 85.

$$3x + 17x + 5 = 3(4) + 17(4) + 5$$
$$= 12 + 68 + 5$$
$$= 80 + 5$$
$$= 85$$

20. The correct answer is 144.

$$2z \times 4 = (2 \times 18) \times 4$$
$$= 36 \times 4$$
$$= 144$$

21. The correct answer is 9.

$$(5x + 2y) \div 3 = [5(3) + 2(6)] \div 3$$
$$= (15 + 12) \div 3$$
$$= 27 \div 3$$
$$= 9$$

22. The correct answer is 80.

$$5m + 3(m + 4n) = 5(7) + 3[(7) + 4(2)]$$
$$= 35 + 3[(7) + 4(2)]$$
$$= 35 + 3(7 + 8)$$
$$= 35 + 3(15)$$
$$= 35 + 45$$
$$= 80$$

23. The correct answer is 320.

$$8(r + 6s) = 8[(4) + 6(6)]$$
$$= 8(4 + 36)$$
$$= 8(40)$$
$$= 320$$

24. The correct answer is 29.

$$(12y + 16z) \div 4 = [12(3) + 16(5)] \div 4$$
$$= (36 + 80) \div 4$$
$$= 116 \div 4$$
$$= 29$$

25. The correct answer is 8.

$$5a - 9b = 5(7) - 9(3)$$
$$= 35 - 27$$
$$= 8$$

Chapter 3

Equations

In this chapter, we'll review the following concepts:

What is an equation?
Parts of equations
Simplifying equations
Solving equations

What is an equation?

An **equation** is a true math statement that contains an equal sign.

Equations can come in several forms.

Some equations contain no variables at all.

$$2 + 2 = 4$$

Other equations contain variables on one side of the equal sign:

$$3x + 4x = 7$$

More complex equations contain variables on both sides of the equal sign:

$$12b + 4b - 9 = 2b + 7b + 19$$

Your job is to find the missing value that makes the statement true.

Parts of equations

Algebra equations contain expressions, equal signs, and sometimes a single number or variable.

$$3x + 4x = 7$$

In the equation shown, $3x + 4x$ is an expression. The number 7 is a single constant.

This equation contains two expressions joined by an equal sign:

$$12b + 4b - 9 = 2b + 7b + 19$$

The two expressions are $12b + 4b - 9$ and $2b + 7b + 19$. In the first expression, on the left side of the equal sign, the numbers 12 and 4 are coefficients. The letter b is a variable, and -9 is a constant. In the second expression, on the right hand side, the coefficients are 2 and 7. The variable is also b, and the number 19 is a constant.

Simplifying equations

Before we can solve equations, we first have to simplify them. This involves three steps:

1. Distribute.

2. Combine like terms.

3. Get all variables on one side of the equation.

Let's review some examples.

Examples

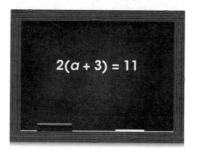

$$2(a + 3) = 11$$

Follow the three steps to simplify:

 Distribute the 2 on the left-hand side.

$$2(a + 3) = 11$$
$$(2 \times a) + (2 \times 3) = 11$$
$$2a + 6 = 11$$

2 Combine like terms.

In this case, we have a 6 on the left side of the equal sign, and an 11 on the right side. These are both constants, so they can be combined. Subtract 6 from both sides of the equation:

$$2a + 6 = 11$$
$$2a + 6 - 6 = 11 - 6$$
$$2a + 0 = 11 - 6$$
$$2a = 5$$

This gives us all like terms combined.

It's fine to add or subtract a number from one side of an equation, as long as you perform the same operation on the other side. This keeps the equation balanced.

3 Get all variables on one side of the equation.

That step is already done! This equation is ready to solve.

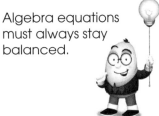

Algebra equations must always stay balanced.

Here's another example with no distribution.

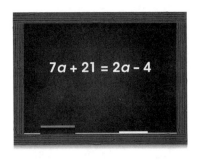

$$7a + 21 = 2a - 4$$

There are no parentheses, so distribution is not necessary. Go to the next step: combine like terms.

First, move all variables to the left side of the equation. The term $7a$ is already on the left, so we must move $2a$. Subtract $2a$ from both sides:

$$7a + 21 = 2a - 4$$
$$7a - 2a + 21 = 2a - 2a - 4$$
$$5a + 21 = -4$$

Next, move all constants to the right side of the equation. Subtract 21 from both sides:

$$5a + 21 = -4$$
$$5a + 21 - 21 = -4 - 21$$
$$5a = -25$$

This equation is ready to solve.

Simplifying using multiplication

You may see some equations with division, as shown:

$$\frac{13q + 7}{11} = 3$$

Even though this equation is more complex than the ones we simplified above, we still simplify it the same way.

1 Distribute.

There are no parentheses here, so skip to the next step.

2 Combine like terms.

In order to combine like terms, we must first remove the 11 from the fraction. To do this, we multiply 11 by both sides:

$$\frac{13q + 7}{11} = 3$$

$$11 \times \left(\frac{13q + 7}{11} \right) = 3 \times 11$$

The 11's cancel out, so we eliminate the fraction:

$$11 \times \left(\frac{13q + 7}{11} \right) = 3 \times 11$$

$$13q + 7 = 3 \times 11$$

$$13q + 7 = 33$$

Now it's much simpler!

We still have one more step to go in combining like terms. Subtract the 7 from both sides:

$$13q + 7 = 33$$

$$13q + 7 - 7 = 33 - 7$$

$$13q = 33 - 7$$

$$13q = 26$$

This one's ready to solve.

Practice Questions—Simplifying equations

Directions: Simplify the following equations. You will find the Practice Question Solutions on page 59.

1. $9z - 32 = z + 16$

2. $25(r + 4) = 20(r + 10)$

3. $17(2b + 9) = 5(6b - 3)$

4. $3v + 4v + 5v = 14v - 24$

5. $\dfrac{7g + 5}{5} = 15$

6. $\dfrac{2w + 4}{2} = 60$

7. $\dfrac{22y + 6}{20} = 2y + 3$

8. $\dfrac{14 + 27n}{3} = n + 7$

9. $32 + \dfrac{5}{6}t = 42$

10. $27 + \dfrac{4}{9}m = 63$

Solving equations

Now that we've learned how to simplify equations, we can work on solving them. We'll start with equations with one variable.

Equations with one variable

To solve equations, as we have seen, the first step is to simplify them. Sometimes simplifying by itself solves the problem.

$$x + 5 = 7$$

To simplify this equation, subtract 5 from both sides:

$$x + 5 = 7$$
$$x + 5 - 5 = 7 - 5$$
$$x = 7 - 5$$
$$x = 2$$

Problem solved!

For most equations, solving requires a few more steps.

Once an equation is in its simplest form, the goal is to have the variable alone on one side of the equation. This is called **isolating the variable**.

In the equation we just solved, the variable x is alone on the left side of the equation. The number 2 is on the right, so the value of x is 2.

But what if we have an equation in its simplest form that looks like this?

$$4z = 24$$

In this case, $4z$ is on one side of the equation, but we want to have z alone. This requires performing the *opposite* of the operation shown.

The values of 4 and z are multiplied, so we must perform division to isolate the z. Divide both sides of the equation by 4:

$$4z = 24$$
$$\frac{4z}{4} = \frac{24}{4}$$

On the left, the 4's cancel out. This leaves us with just z:

$$\frac{4z}{4} = \frac{24}{4}$$
$$z = \frac{24}{4}$$

Now, divide 24 by 4 to get the answer. The value of z is 6.

$$\frac{4z}{4} = \frac{24}{4}$$
$$z = 6$$

Practice Questions—Equations with one variable

Directions: Solve each of the following equations. You will find the Practice Question Solutions on page 61.

11. If $7a = 14$, find the value of a.

12. If $3b = 18$, find the value of b.

13. If $5e = 45$, find the value of e.

14. If $10z = 1,000$, find the value of z.

15. What is the value of k if $7k = 365$?

16. If $3d + 2 = 8$, find the value of d.

17. What is the value of p if $8p - 5 = 11$?

18. If $10e + 5 = 85$, solve for e.

19. What is the value of q if $6q - 10 = 98$?

20. If $9r + 1 = 19$, find the value of r.

egghead's Guide to Algebra

Equations with variables on both sides

Even if an equation has variables on both sides, the process of solving is the same. Simplify first, then isolate the variable. Then perform the opposite operation.

Examples

$$3 - x = x + 1$$

In this case, there is no distribution to perform, so we combine like terms:

$$3 - x - x = x - x + 1$$
$$3 - 2x = 1$$
$$3 - 3 - 2x = 1 - 3$$
$$-2x = 1 - 3$$
$$-2x = -2$$

Next, isolate the variable:

$$-2x = -2$$
$$\frac{-2x}{-2} = \frac{-2}{-2}$$
$$x = 1$$

The value of x is 1.

Practice Questions—Equations with variables on both sides

Directions: Solve each of the following equations. You will find the Practice Question Solutions on page 63.

21. $2r - 3 = r$

22. $x + 12 = 13x$

23. $10y + 3 = 3y - 4$

24. $a + 17a = 40 - 2a$

25. $5z + 25 = z + 1$

26. $2n + 7n + 16 = 6n - 20$

27. $3(c + 4) = 12(c + 1)$

28. $24 + 2k = 7(3 - k)$

29. $8(7m + 2) = 8(3m + 4)$

30. $5g + 14 = 6(2g - 7)$

Chapter Review

Solutions can be found on page 66.

1. If $10c = 20$, find the value of c.

2. What is the value of w if $13w = 117$?

3. If $2x = 22$, find the value of x.

4. If $5f = 125$, solve for f.

5. What is the value of g if $7g = 28$?

6. If $4s + 8 = 20$, find the value of s.

7. What is the value of y if $8y - 5 = 35$?

8.

If $h + 100 = 200$, solve for h.

9. What is the value of v if $2v + 11 = 15$?

10. If $6a - 9 = 81$, find the value of a.

11. If $7a + 21 = 2a - 4$, find the value of a.

12. If $9z - 32 = z + 16$, solve for z.

13. If $25(r + 4) = 20(r + 10)$, find the value of r.

14. If $17(2b + 9) = 5(6b - 3)$, solve for b.

15. What is the value of *v* if $3v + 4v + 5v = 14v - 24$?

16. If $\dfrac{13q + 7}{11} = 3$, find the value of *q*.

17. What is the value of *g* if $\dfrac{7g + 5}{5} = 15$?

18. What is the value of *w* if $\dfrac{2w + 4}{2} = 60$?

19. What is the value of *y* if $\dfrac{22y + 6}{20} = 2y + 3$?

20. If $\dfrac{14 + 27n}{3} = n + 7$, find the value of *n*.

21.

If $\dfrac{1}{3}p = 3$, solve for *p*.

22. What is the value of *d* if $\dfrac{3}{4}d = 60$?

23. If $6 + \dfrac{1}{2}s = 14$, find the value of *s*.

24. What is the value of *t* if $32 + \dfrac{5}{6}t = 42$?

25. What is the value of *m* if $27 + \dfrac{4}{9}m = 63$?

Practice Question Solutions

Simplifying equations

1. The simplified form of the equation is $8z = 48$.

$$9z - 32 = z + 16$$
$$9z - z - 32 = z - z + 16$$
$$8z - 32 = 16$$
$$8z - 32 + 32 = 16 + 32$$
$$8z = 48$$

2. The simplified form of the equation is $5r = 100$.

$$25(r + 4) = 20(r + 10)$$
$$(25 \times r) + (25 \times 4) = (20 \times r) + (20 \times 10)$$
$$25r + 100 = 20r + 200$$
$$25r - 20r + 100 = 20r - 20r + 200$$
$$5r + 100 = 200$$
$$5r + 100 - 100 = 200 - 100$$
$$5r = 100$$

3.

The simplified form of the equation is $4b = -168$.

$$17(2b + 9) = 5(6b - 3)$$
$$(17 \times 2b) + (17 \times 9) = (5 \times 6b) - (5 \times 3)$$
$$34b + 153 = 30b - 15$$
$$34b - 30b + 153 = 30b - 30b - 15$$
$$4b + 153 = -15$$
$$4b + 153 - 153 = -15 - 153$$
$$4b = -168$$

4. The simplified form of the equation is $-2v = -24$.

$$3v + 4v + 5v = 14v - 24$$
$$12v = 14v - 24$$
$$12v - 14v = 14v - 14v - 24$$
$$-2v = -24$$

5. The simplified form of the equation is $7g = 70$.

There is no distribution here, but we must remove the number 5 from the fraction, as we saw in the example above. Multiply 5 by both sides:

$$\frac{7g + 5}{5} = 15$$

$$5 \times \left(\frac{7g + 5}{5}\right) = 15 \times 5$$

$$7g + 5 = 15 \times 5$$

$$7g + 5 = 75$$

$$7g + 5 - 5 = 75 - 5$$

$$7g = 70$$

6. The simplified form of the equation is $2w = 116$.

$$\frac{2w + 4}{2} = 60$$

$$2 \times \left(\frac{2w + 4}{2}\right) = 60 \times 2$$

$$2w + 4 = 60 \times 2$$

$$2w + 4 = 120$$

$$2w + 4 - 4 = 120 - 4$$

$$2w = 116$$

7. The simplified form of the equation is $-18y = 54$.

$$\frac{22y + 6}{20} = 2y + 3$$

$$20\left(\frac{22y + 6}{20}\right) = 20(2y + 3)$$

$$22y + 6 = 20(2y + 3)$$

$$22y + 6 = (20 \times 2y) + (20 \times 3)$$

$$22y + 6 = 40y + 60$$

$$22y - 40y + 6 = 40y - 40y + 60$$

$$22y - 40y + 6 = 60$$

$$-18y + 6 = 60$$

$$-18y + 6 - 6 = 60 - 6$$

$$-18y = 54$$

8. The simplified form of the equation is $24n = 7$.

$$\frac{14 + 27n}{3} = n + 7$$

$$3 \times \left(\frac{14 + 27n}{3}\right) = 3(n + 7)$$

$$14 + 27n = 3(n + 7)$$

$$14 + 27n = (3 \times n) + (3 \times 7)$$

$$14 + 27n = 3n + 21$$

$$14 + 27n - 3n = 3n - 3n + 21$$

$$14 + 24n = 21$$

$$14 - 14 + 24n = 21 - 14$$

$$24n = 7$$

9. The simplified form of the equation is

$$\frac{5}{6}t = 10 \, .$$

In this problem, the coefficient of t is a fraction. We still follow the same three steps to simplify:

1 Distribute.

2 Combine like terms.

3 Get all variables on one side of the equation.

There is no need for distribution, so we skip to step 2.

Subtract the number 32 from both sides.

$$32 + \frac{5}{6}t = 42$$

$$32 - 32 + \frac{5}{6}t = 42 - 32$$

$$\frac{5}{6}t = 42 - 32$$

$$\frac{5}{6}t = 10$$

This leaves a simplified equation of

$$\frac{5}{6}t = 10.$$

10. The simplified form of the equation is

$$\frac{4}{9}m = 36.$$

Subtract 27 from both sides:

$$27 + \frac{4}{9}m = 63$$

$$27 - 27 + \frac{4}{9}m = 63 - 27$$

$$\frac{4}{9}m = 63 - 27$$

$$\frac{4}{9}m = 36$$

Solving equations with one variable

11. The value of a is 2.

This equation is already in its simplest form, so we divide both sides by 7:

$$7a = 14$$

$$\frac{7a}{7} = \frac{14}{7}$$

$$a = \frac{14}{7}$$

$$a = 2$$

12. The value of b is 6.

$$3b = 18$$

$$\frac{3b}{3} = \frac{18}{3}$$

$$b = \frac{18}{3}$$

$$b = 6$$

13. The value of e is 9.

$$5e = 45$$

$$\frac{5e}{5} = \frac{45}{5}$$

$$e = \frac{45}{5}$$

$$e = 9$$

14. The value of z is 100.

$$10z = 1,000$$

$$\frac{10z}{10} = \frac{1,000}{10}$$

$$z = \frac{1,000}{10}$$

$$z = 100$$

15. The value of k is approximately 52.14.

$$7k = 365$$

$$\frac{7k}{7} = \frac{365}{7}$$

$$k = \frac{365}{7}$$

$$k \approx 52.14$$

16. The value of d is 2.

To solve, first simplify the equation.
Subtract 2 from both sides:

$$3d + 2 = 8$$
$$3d + 2 - 2 = 8 - 2$$
$$3d = 8 - 2$$
$$3d = 6$$

Next, divide both sides by 3.

$$\frac{3d}{3} = \frac{6}{3}$$
$$d = \frac{6}{3}$$
$$d = 2$$

17. The value of p is 2.

$$8p - 5 = 11$$
$$8p - 5 + 5 = 11 + 5$$
$$8p = 11 + 5$$
$$8p = 16$$
$$\frac{8p}{8} = \frac{16}{8}$$
$$p = \frac{16}{8}$$
$$p = 2$$

18. The value of e is 8.

$$10e + 5 = 85$$
$$10e + 5 - 5 = 85 - 5$$
$$10e = 85 - 5$$
$$10e = 80$$
$$\frac{10e}{10} = \frac{80}{10}$$
$$e = \frac{80}{10}$$
$$e = 8$$

19. The value of q is 18.

$$6q - 10 = 98$$
$$6q - 10 + 10 = 98 + 10$$
$$6q = 98 + 10$$
$$6q = 108$$
$$\frac{6q}{6} = \frac{108}{6}$$
$$q = \frac{108}{6}$$
$$q = 18$$

20. The value of r is 2.

$$9r + 1 = 19$$
$$9r + 1 - 1 = 19 - 1$$
$$9r = 19 - 1$$
$$9r = 18$$
$$\frac{9r}{9} = \frac{18}{9}$$
$$r = \frac{18}{9}$$
$$r = 2$$

Solving equations with variables on both sides

21. The value of r is 3.

$$2r - 3 = r$$
$$2r - 3 + 3 = r + 3$$
$$2r = r + 3$$
$$2r - r = r - r + 3$$
$$r = 3$$

22. The value of x is 1.

$$x + 12 = 13x$$
$$x + 12 - 12 = 13x - 12$$
$$x = 13x - 12$$
$$x - 13x = 13x - 13x - 12$$
$$x - 13x = -12$$
$$-12x = -12$$
$$\frac{-12x}{-12} = \frac{-12}{-12}$$
$$x = \frac{-12}{-12}$$
$$x = 1$$

23. The value of y is –1.

$$10y + 3 = 3y - 4$$
$$10y + 3 - 3 = 3y - 4 - 3$$
$$10y = 3y - 4 - 3$$
$$10y = 3y - 7$$
$$10y - 3y = 3y - 3y - 7$$
$$10y - 3y = -7$$
$$7y = -7$$
$$\frac{7y}{7} = \frac{-7}{7}$$
$$y = \frac{-7}{7}$$
$$y = -1$$

24.

The value of a is 2.

$$a + 17a = 40 - 2a$$
$$a + 17a + 2a = 40 - 2a + 2a$$
$$a + 17a + 2a = 40$$
$$20a = 40$$
$$\frac{20a}{20} = \frac{40}{20}$$
$$a = \frac{40}{20}$$
$$a = 2$$

25. The value of z is –6.

$$5z + 25 = z + 1$$
$$5z - z + 25 = z - z + 1$$
$$5z - z + 25 = 1$$
$$4z + 25 = 1$$
$$4z + 25 - 25 = 1 - 25$$
$$4z = 1 - 25$$
$$4z = -24$$
$$\frac{4z}{4} = \frac{-24}{4}$$
$$z = \frac{-24}{4}$$
$$z = -6$$

26. The value of n is -12.

$$2n + 7n + 16 = 6n - 20$$
$$2n + 7n - 6n + 16 = 6n - 6n - 20$$
$$2n + 7n - 6n + 16 = -20$$
$$3n + 16 = -20$$
$$3n + 16 - 16 = -20 - 16$$
$$3n = -20 - 16$$
$$3n = -36$$
$$\frac{3n}{3} = \frac{-36}{3}$$
$$n = \frac{-36}{3}$$
$$n = -12$$

27. The value of c is 0.

$$3(c + 4) = 12(c + 1)$$
$$(3 \times c) + (3 \times 4) = (12 \times c) + (12 \times 1)$$
$$3c + 12 = 12c + 12$$
$$3c + 12 - 12 = 12c + 12 - 12$$
$$3c = 12c + 0$$
$$3c - 12c = 0$$
$$-9c = 0$$
$$c = 0$$

28. The value of k is $-\frac{1}{3}$.

$$24 + 2k = 7(3 - k)$$
$$24 + 2k = (7 \times 3) - (7 \times k)$$
$$24 + 2k = 21 - 7k$$
$$24 - 24 + 2k = 21 - 24 - 7k$$
$$2k = 21 - 24 - 7k$$
$$2k = -3 - 7k$$
$$2k + 7k = -3 - 7k + 7k$$
$$9k = -3$$
$$\frac{9k}{9} = \frac{-3}{9}$$
$$k = \frac{-3}{9}$$
$$k = -\frac{1}{3}$$

29. The value of m is $\frac{1}{2}$.

$$8(7m + 2) = 8(3m + 4)$$
$$(8 \times 7m) + (8 \times 2) = (8 \times 3m) + (8 \times 4)$$
$$56m + 16 = 24m + 32$$
$$56m + 16 - 16 = 24m + 32 - 16$$
$$56m = 24m + 32 - 16$$
$$56m = 24m + 16$$
$$56m - 24m = 24m - 24m + 16$$
$$56m - 24m = 16$$
$$32m = 16$$
$$\frac{32m}{32} = \frac{16}{32}$$
$$m = \frac{16}{32}$$
$$m = \frac{1}{2}$$

30. The value of g is 8.

$$5g + 14 = 6(2g - 7)$$
$$5g + 14 = (6 \times 2g) - (6 \times 7)$$
$$5g + 14 = 12g - 42$$
$$5g + 14 - 14 = 12g - 42 - 14$$
$$5g = 12g - 42 - 14$$
$$5g = 12g - 56$$
$$5g - 12g = 12g - 12g - 56$$
$$5g - 12g = -56$$
$$-7g = -56$$
$$\frac{-7g}{-7} = \frac{-56}{-7}$$
$$g = \frac{-56}{-7}$$
$$g = 8$$

Great work!

Chapter Review Solutions

1. The value of *c* is 2.

$$10c = 20$$
$$\frac{10c}{10} = \frac{20}{10}$$
$$c = \frac{20}{10}$$
$$c = 2$$

2. The value of *w* is 9.

$$13w = 117$$
$$\frac{13w}{13} = \frac{117}{13}$$
$$w = \frac{117}{13}$$
$$w = 9$$

3. The value of *x* is 11.

$$2x = 22$$
$$\frac{2x}{2} = \frac{22}{2}$$
$$x = \frac{22}{2}$$
$$x = 11$$

4. The value of *f* is 25.

$$5f = 125$$
$$\frac{5f}{5} = \frac{125}{5}$$
$$f = \frac{125}{5}$$
$$f = 25$$

5. The value of *g* is 4.

$$7g = 28$$
$$\frac{7g}{7} = \frac{28}{7}$$
$$g = \frac{28}{7}$$
$$g = 4$$

6.

The value of *s* is 3.

$$4s + 8 = 20$$
$$4s + 8 - 8 = 20 - 8$$
$$4s = 20 - 8$$
$$4s = 12$$
$$\frac{4s}{4} = \frac{12}{4}$$
$$s = \frac{12}{4}$$
$$s = 3$$

7. The value of *y* is 5.

$$8y - 5 = 35$$
$$8y - 5 + 5 = 35 + 5$$
$$8y = 35 + 5$$
$$8y = 40$$
$$\frac{8y}{8} = \frac{40}{8}$$
$$y = \frac{40}{8}$$
$$y = 5$$

8. The value of h is 100.

$$h + 100 = 200$$
$$h + 100 - 100 = 200 - 100$$
$$h = 200 - 100$$
$$h = 100$$

9. The value of v is 2.

$$2v + 11 = 15$$
$$2v + 11 - 11 = 15 - 11$$
$$2v = 15 - 11$$
$$2v = 4$$
$$\frac{2v}{2} = \frac{4}{2}$$
$$v = \frac{4}{2}$$
$$v = 2$$

10. The value of a is 15.

$$6a - 9 = 81$$
$$6a - 9 + 9 = 81 + 9$$
$$6a = 81 + 9$$
$$6a = 90$$
$$\frac{6a}{6} = \frac{90}{6}$$
$$a = \frac{90}{6}$$
$$a = 15$$

11.

The value of a is −5.

$$7a + 21 = 2a - 4$$
$$7a - 2a + 21 = 2a - 2a - 4$$
$$5a + 21 = -4$$
$$5a + 21 - 21 = -4 - 21$$
$$5a = -25$$
$$\frac{5a}{5} = \frac{-25}{5}$$
$$a = \frac{-25}{5}$$
$$a = -5$$

12. The value of z is 6.

$$9z - 32 = z + 16$$
$$9z - z - 32 = z - z + 16$$
$$8z - 32 = 16$$
$$8z - 32 + 32 = 16 + 32$$
$$8z = 48$$
$$\frac{8z}{8} = \frac{48}{8}$$
$$z = \frac{48}{8}$$
$$z = 6$$

13. The value of r is 20.

$$25(r + 4) = 20(r + 10)$$
$$(25 \times r) + (25 \times 4) = (20 \times r) + (20 \times 10)$$
$$25r + 100 = 20r + 200$$
$$25r - 20r + 100 = 20r - 20r + 200$$
$$5r + 100 = 200$$
$$5r + 100 - 100 = 200 - 100$$
$$5r = 100$$
$$\frac{5r}{5} = \frac{100}{5}$$
$$r = \frac{100}{5}$$
$$r = 20$$

14. The value of b is –42.

$$17(2b + 9) = 5(6b - 3)$$
$$(17 \times 2b) + (17 \times 9) = (5 \times 6b) - (5 \times 3)$$
$$34b + 153 = 30b - 15$$
$$34b - 30b + 153 = 30b - 30b - 15$$
$$4b + 153 = -15$$
$$4b + 153 - 153 = -15 - 153$$
$$4b = -168$$
$$\frac{4b}{4} = \frac{-168}{4}$$
$$b = \frac{-168}{4}$$
$$b = -42$$

15. The value of v is 12.

$$3v + 4v + 5v = 14v - 24$$
$$12v = 14v - 24$$
$$12v - 14v = 14v - 14v - 24$$
$$-2v = -24$$
$$\frac{-2v}{-2} = \frac{-24}{-2}$$
$$v = \frac{-24}{-2}$$
$$v = 12$$

16. The value of q is 2.

We saw this equation simplified in the question.

$$\frac{13q + 7}{11} = 3$$
$$11 \times \left(\frac{13q + 7}{11}\right) = 3 \times 11$$
$$13q + 7 = 33$$
$$13q + 7 - 7 = 33 - 7$$
$$13q = 26$$
$$\frac{13q}{13} = \frac{26}{13}$$
$$q = \frac{26}{13}$$
$$q = 2$$

17. The value of g is 10.

$$\frac{7g + 5}{5} = 15$$
$$5 \times \left(\frac{7g + 5}{5}\right) = 15 \times 5$$
$$7g + 5 = 15 \times 5$$
$$7g + 5 = 75$$
$$7g + 5 - 5 = 75 - 5$$
$$7g = 70$$
$$\frac{7g}{7} = \frac{70}{7}$$
$$g = \frac{70}{7}$$
$$g = 10$$

18. The value of w is 58.

$$\frac{2w+4}{2} = 60$$
$$2 \times \left(\frac{2w+4}{2}\right) = 60 \times 2$$
$$2w + 4 = 60 \times 2$$
$$2w + 4 = 120$$
$$2w + 4 - 4 = 120 - 4$$
$$2w = 116$$
$$\frac{2w}{2} = \frac{116}{2}$$
$$w = \frac{116}{2}$$
$$w = 58$$

19. The value of y is -3.

$$\frac{22y+6}{20} = 2y + 3$$
$$20\left(\frac{22y+6}{20}\right) = 20(2y+3)$$
$$22y + 6 = 20(2y+3)$$
$$22y + 6 = (20 \times 2y) + (20 \times 3)$$
$$22y + 6 = 40y + 60$$
$$22y - 40y + 6 = 40y - 40y + 60$$
$$22y - 40y + 6 = 60$$
$$-18y + 6 = 60$$
$$-18y + 6 - 6 = 60 - 6$$
$$-18y = 54$$
$$\frac{-18y}{-18} = \frac{54}{-18}$$
$$y = \frac{54}{-18}$$
$$y = -3$$

20. The value of n is $\frac{7}{24}$.

$$\frac{14+27n}{3} = n + 7$$
$$3 \times \left(\frac{14+27n}{3}\right) = 3(n+7)$$
$$14 + 27n = 3(n+7)$$
$$14 + 27n = (3 \times n) + (3 \times 7)$$
$$14 + 27n = 3n + 21$$
$$14 + 27n - 3n = 3n - 3n + 21$$
$$14 + 24n = 21$$
$$14 - 14 + 24n = 21 - 14$$
$$24n = 7$$
$$\frac{24n}{24} = \frac{7}{24}$$
$$n = \frac{7}{24}$$

In this case, when we divide both sides of the equation by 24, we get a fraction.

The fraction $\frac{7}{24}$ can't be reduced, so the value of n is $\frac{7}{24}$.

21. The value of p is 9.

To remove the fraction on the left-hand side, multiply both sides of the equation by 3:

$$\frac{1}{3}p = 3$$
$$3 \times \left(\frac{1}{3}p\right) = 3 \times 3$$
$$p = 3 \times 3$$
$$p = 9$$

The 3's on the left cancel out, so the correct answer is 9.

Chapter 3: Equations

22. The value of *d* is 80.

Multiply both sides of the equation by 4:

$$\frac{3}{4}d = 60$$

$$4 \times \left(\frac{3}{4}d\right) = 60 \times 4$$

$$3d = 240$$

This removes the fraction.

Divide both sides by 3 to solve.

$$3d = 240$$

$$\frac{3d}{3} = \frac{240}{3}$$

$$d = \frac{240}{3}$$

$$d = 80$$

23. The value of *s* is 16.

$$6 + \frac{1}{2}s = 14$$

$$6 - 6 + \frac{1}{2}s = 14 - 6$$

$$\frac{1}{2}s = 14 - 6$$

$$\frac{1}{2}s = 8$$

$$2 \times \left(\frac{1}{2}s\right) = 8 \times 2$$

$$s = 8 \times 2$$

$$s = 16$$

24. The value of *t* is 12.

$$32 + \frac{5}{6}t = 42$$

$$32 - 32 + \frac{5}{6}t = 42 - 32$$

$$\frac{5}{6}t = 42 - 32$$

$$\frac{5}{6}t = 10$$

$$6 \times \left(\frac{5}{6}t\right) = 10 \times 6$$

$$5t = 10 \times 6$$

$$5t = 60$$

$$\frac{5t}{5} = \frac{60}{5}$$

$$t = 12$$

25. The value of *m* is 81.

$$27 + \frac{4}{9}m = 63$$

$$27 - 27 + \frac{4}{9}m = 63 - 27$$

$$\frac{4}{9}m = 63 - 27$$

$$\frac{4}{9}m = 36$$

$$9 \times \left(\frac{4}{9}m\right) = 36 \times 9$$

$$4m = 36 \times 9$$

$$4m = 324$$

$$\frac{4m}{4} = \frac{324}{4}$$

$$m = \frac{324}{4}$$

$$m = 81$$

Chapter 4

Exponents

In this chapter, we'll review the following concepts:

What is an exponent?
Exponents of positive and negative numbers
Operations with exponents
Negative exponents

What is an exponent?

An **exponent** is a mathematical notation that indicates how many times a number is multiplied by itself.

An exponent is written as a small number next to a larger number, known as the base:

In this example, the number 2 is the exponent. The number 4 is the base.

This notation tells us to multiply the number 4 by itself:

$$4^2 = 4 \times 4$$

The notation 10^5 tells us that 10 is being multiplied five times:

$$10^5 = 10 \times 10 \times 10 \times 10 \times 10$$

To calculate the value of a number raised to an exponent, simply multiply the number by itself that many times. Here are some examples.

$$2^3 = 2 \times 2 \times 2$$

$$5^4 = 5 \times 5 \times 5 \times 5$$

$$a^2 = a \times a$$

There are some special cases of exponents that come up often in algebra. Any number with an exponent of 0 equals 1:

$$2^0 = 1$$

$$19^0 = 1$$

$$n^0 = 1$$

Any number with an exponent of 1 is equal to itself:

$$5^1 = 5$$

$$120^1 = 120$$

$$a^1 = a$$

Any number multiplied by itself is known as a **perfect square**.

$$3^2 = 3 \times 3$$

$$42^2 = 42 \times 42$$

$$m^2 = m \times m$$

We also refer to the expression 3^2 as "three squared." The variable x with an exponent of 2 would be "x squared."

Any number multiplied by itself three times is known as a **perfect cube**.

$$7^3 = 7 \times 7 \times 7$$

$$9^3 = 9 \times 9 \times 9$$

$$z^3 = z \times z \times z$$

The expression 7^3 is referred to as "seven cubed." The variable c with an exponent of 3 is "c cubed."

Exponents of positive and negative numbers

It doesn't matter how many times you multiply a positive number by itself—the result is always positive.

For example, $12 \times 12 = 144$.

With negative numbers, the results can be positive or negative.

The value of $(-2)^2$ is positive: $-2 \times -2 = 4$.

The value of $(-2)^3$ is negative: $-2 \times -2 \times -2 = -8$

The result depends on whether the exponent is even or odd. If a negative base is raised to an even exponent, the resulting number will be positive. If a negative base is raised to an odd exponent, the resulting number will be negative.

Base	Exponent	Result	Example
+	even	+	$2^2 = 4$
+	odd	+	$2^3 = 8$
–	even	+	$(-2)^2 = 4$
–	odd	–	$(-2)^3 = -8$

Here's a chart to help you keep track.

Operations with exponents

When a number is raised to an exponent, we might also say that number is raised to a certain **power**. So, 4^2 could be called "four squared" or "four to the second power." The expression 8^3 could be "eight cubed" or "eight to the third power," and so on.

In the first chapter, Working with Variables, we reviewed the order of operations using PEMDAS. When we solve an algebra problem, we perform operations in parentheses first. Operations with exponents are performed second. Here are some typical operations with exponents and how they should be performed.

When **multiplying** exponential expressions with the same base, the exponents are added:

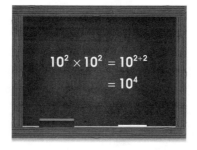

$$10^2 \times 10^2 = 10^{2+2}$$
$$= 10^4$$

When **dividing** exponential expressions with the same base, the exponents are subtracted:

$$\frac{8^4}{8^1} = 8^{4-1}$$
$$= 8^3$$

The bases must be the same for this to work.

When **raising a power to another power**, the exponents are multiplied:

$$\left(6^3\right)^2 = 6^{3\times2}$$
$$= 6^6$$

When **raising a product to a power**, each number is taken to that power:

$$\left(7\times2\right)^5 = 7^5 \times 2^5$$

When **raising a quotient to a power**, each number is taken to that power as well:

$$\left(\frac{19}{3}\right)^4 = \frac{19^4}{3^4}$$

Now try some practice questions.

Practice Questions—Operations with exponents

Directions: Perform the operations shown. Leave your answers in exponential form. You will find the Practice Question Solutions on page 83.

1. $9^2 \times 9^4$

2. $\dfrac{7^4}{7^2}$

3. $\left(25^3\right)^2$

4. $(4 \times 12)^3$

5. $\left(\dfrac{20}{4}\right)^5$

6. $a^6 \times a^3$

7. $\dfrac{k^7}{k^3}$

8. $\left(n^4\right)^9$

9. $(x \times y)^3$

10. $\left(\dfrac{r}{s}\right)^4$

Negative exponents

So far we have worked just with positive exponents. But exponents can be negative as well:

When we have a number raised to a negative power, we take the reciprocal of that number raised to the opposite power. The base becomes a fraction, and the negative exponent becomes positive:

$$4^{-2} = \frac{1}{4^2}$$

This can also be written as:

$$\left(\frac{1}{4}\right)^2$$

Here are more examples.

$$7^{-3} = \frac{1}{7^3}$$
$$12^{-n} = \frac{1}{12^n}$$
$$r^{-2} = \frac{1}{r^2}$$

egghead's Guide to Algebra

Practice Questions—Negative exponents

Directions: Perform the operations shown. Leave your answers in exponential form. You will find the Practice Question Solutions on page 84.

11. 5^{-4}

12. 3^{-3}

13. 6^{-7}

14. 15^{-n}

15. 13^{-p}

16. 21^{-s}

17. x^{-2}

18. z^{-4}

19. a^{-b}

20. r^{-q}

Chapter 4: Exponents

These are the basic rules for working with exponents.

Here's a chart that summarizes them.

Expression	Operation	Result
$n^a \times n^b$	Add exponents	n^{a+b}
$\dfrac{n^a}{n^b}$	Subtract exponents	n^{a-b}
$(n^a)^b$	Multiply exponents	$n^{a \times b}$
$(n \times m)^a$	Raise bases to that power	$n^a \times m^a$
$\left(\dfrac{n}{m}\right)^a$	Raise bases to that power	$\dfrac{n^a}{m^a}$
n^{-a}	Take reciprocal of base with positive exponent	$\dfrac{1}{n^a}$

Chapter Review

Directions: Perform the operations shown. Leave your answers in exponential form. Solutions can be found on page 85.

1. $6^3 \times 6^7$

2. $d^5 \times d^4$

3. $g^{10} \times g^{11}$

4. $z^2 \times z^9$

5. $\dfrac{9^{12}}{9^3}$

6. $\dfrac{18^5}{18^2}$

7. $\dfrac{n^{12}}{n^2}$

8. $\dfrac{q^{14}}{q^9}$

9. $\left(14^7\right)^5$

10. $\left(2^4\right)^{10}$

11. $\left(60^{20}\right)^3$

12. $\left(f^6\right)^2$

13. $(9 \times 2)^2$

14. $(15 \times 7)^5$

15. $(h \times j)^7$

16. $(n \times p)^6$

17. $\left(\dfrac{100}{7}\right)^6$

18. $\left(\dfrac{33}{10}\right)^2$

19. $\left(\dfrac{y}{p}\right)^3$

20. $\left(\dfrac{c}{d}\right)^7$

21. 13^{-2}

22. $e^{\,3}$

23. u^{-7}

24. s^{-6}

25. p^{-z}

Practice Question Solutions

Operations with exponents

1. The correct answer is 9^6.

   Here, two exponential expressions with the same base are being multiplied. Therefore, the exponents are added:

   $$9^2 \times 9^4 = 9^{2+4}$$
   $$= 9^6$$

2. The correct answer is 7^2.

   This time, exponential expressions with the same base are being divided. So, we subtract the exponents:

   $$\frac{7^4}{7^2} = 7^{4-2}$$
   $$= 7^2$$

3. The correct answer is 25^6.

   In this operation, we raise a power to another power. This means the exponents are multiplied:

   $$\left(25^3\right)^2 = 25^{3\times2}$$
   $$= 25^6$$

4. The correct answer is $4^3 \times 12^3$.

   This expression contains a product raised to a power. So, we take each number to that power.

   $$(4 \times 12)^3 = 4^3 \times 12^3$$

5. The correct answer is $\dfrac{20^5}{4^5}$.

   Here, we are raising a quotient to a power. Like with products raised to a power, each number must be taken to that power:

   $$\left(\frac{20}{4}\right)^5 = \frac{20^5}{4^5}$$

6. The correct answer is a^9.

   Two exponential expressions with the same base are being multiplied. Add the exponents:

   $$a^6 \times a^3 = a^{6+3}$$
   $$= a^9$$

7. The correct answer is k^4.

   Two exponential expressions with the same base are being divided, so the exponents are subtracted:

   $$\frac{k^7}{k^3} = k^{7-3}$$
   $$= k^4$$

Nice work!

8. The correct answer is n^{36}.

This question contains a power, n^4, being raised to another power. The exponents are multiplied:

$$\left(n^4\right)^9 = n^{4\times9}$$
$$= n^{36}$$

9. The correct answer is $x^3 \times y^3$. It could also be written x^3y^3.

This problem contains a product raised to a power. So, each number is taken to that power:

$$(x \times y)^3 = x^3 \times y^3$$

10. The correct answer is $\dfrac{r^4}{s^4}$.

Here we have a quotient raised to a power. That means each number is taken to that power:

$$\left(\frac{r}{s}\right)^4 = \frac{r^4}{s^4}$$

Negative exponents

11. The correct answer is $\dfrac{1}{5^4}$.

Take the reciprocal of the base, with a positive exponent:

$$5^{-4} = \frac{1}{5^4}$$

12. The correct answer is $\dfrac{1}{3^3}$.

Take the reciprocal of the base, with a positive exponent:

$$3^{-3} = \frac{1}{3^3}$$

13.

The correct answer is $\dfrac{1}{6^7}$.

Take the reciprocal of the base, with a positive exponent:

$$6^{-7} = \frac{1}{6^7}$$

14. The correct answer is $\dfrac{1}{15^n}$.

Take the reciprocal of the base, with a positive exponent:

$$15^{-n} = \frac{1}{15^n}$$

15. The correct answer is $\dfrac{1}{13^p}$.

Take the reciprocal of the base, with a positive exponent:

$$13^{-p} = \frac{1}{13^p}$$

16. The correct answer is $\dfrac{1}{21^s}$.

$$21^{-s} = \frac{1}{21^s}$$

17. The correct answer is $\dfrac{1}{x^2}$.

$$x^{-2} = \frac{1}{x^2}$$

18. The correct answer is $\dfrac{1}{z^4}$.

$$z^{-4} = \dfrac{1}{z^4}$$

19. The correct answer is $\dfrac{1}{a^b}$.

$$a^{-b} = \dfrac{1}{a^b}$$

20. The correct answer is $\dfrac{1}{r^q}$.

$$r^{-q} = \dfrac{1}{r^q}$$

Chapter Review Solutions

1. The correct answer is 6^{10}.

Two exponential expressions with the same base are being multiplied. So, we add the exponents:

$$6^3 \times 6^7 = 6^{3+7}$$
$$= 6^{10}$$

2. The correct answer is d^9.

Two exponential expressions with the same base are being multiplied. Therefore, the exponents are added:

$$d^5 \times d^4 = d^{5+4}$$
$$= d^9$$

3. The correct answer is g^{21}.

Add the exponents:

$$g^{10} \times g^{11} = g^{10+11}$$
$$= g^{21}$$

4. The correct answer is z^{11}.

The bases are the same, so the exponents are added:

$$z^2 \times z^9 = z^{2+9}$$
$$= z^{11}$$

5. The correct answer is 9^9.

Exponential expressions with the same base are being divided. So, we subtract the exponents:

$$\frac{9^{12}}{9^3} = 9^{12-3}$$
$$= 9^9$$

6. The correct answer is 18^3.

Exponential expressions with the same base are being divided. The exponents are subtracted:

$$\frac{18^5}{18^2} = 18^{5-2}$$
$$= 18^3$$

7. The correct answer is n^{10}.

The bases are the same, so subtract the exponents.

$$\frac{n^{12}}{n^2} = n^{12-2}$$
$$= n^{10}$$

8. The correct answer is q^5.

Exponential expressions with the same base are being divided, so we subtract the exponents:

$$\frac{q^{14}}{q^9} = q^{14-9}$$
$$= q^5$$

9. The correct answer is 14^{35}.

In this operation, we raise a power to another power. This means the exponents are multiplied:

$$\left(14^7\right)^5 = 14^{7\times5}$$
$$= 14^{35}$$

10. The correct answer is 2^{40}.

A power is being raised to another power. So, the exponents are multiplied:

$$\left(2^4\right)^{10} = 2^{4\times10}$$
$$= 2^{40}$$

11. The correct answer is 60^{60}.

Multiply the exponents:

$$\left(60^{20}\right)^3 = 60^{20\times3}$$
$$= 60^{60}$$

12. The correct answer is f^{12}.

A power is being raised to another power. So, the exponents are multiplied:

$$\left(f^6\right)^2 = f^{6\times2}$$
$$= f^{12}$$

13. The correct answer is $9^2 \times 2^2$.

This expression contains a product raised to a power. Take each number to that power:

$$(9 \times 2)^2 = 9^2 \times 2^2$$

14. The correct answer is $15^5 \times 7^5$.

Take each number to the power of 5.

$(15 \times 7)^5 = 15^5 \times 7^5$

15. The correct answer is $h^7 \times j^7$.

Take each number to the power of 7:

$(h \times j)^7 = h^7 \times j^7$

16. The correct answer is $n^6 \times p^6$.

Take each number to the 6th power:

$(n \times p)^6 = n^6 \times p^6$

17. The correct answer is $\dfrac{100^6}{7^6}$.

Here, we are raising a quotient to a power. Like with products raised to a power, each number must be taken to that power:

$\left(\dfrac{100}{7}\right)^6 = \dfrac{100^6}{7^6}$

18. The correct answer is $\dfrac{33^2}{10^2}$.

Take each number to the power of 2:

$\left(\dfrac{33}{10}\right)^2 = \dfrac{33^2}{10^2}$

19. The correct answer is $\dfrac{y^3}{p^3}$.

Take each number to the power of 3:

$\left(\dfrac{y}{p}\right)^3 = \dfrac{y^3}{p^3}$

20. The correct answer is $\dfrac{c^7}{d^7}$.

This is a quotient raised to a power. Take each number to that power:

$\left(\dfrac{c}{d}\right)^7 = \dfrac{c^7}{d^7}$

21. The correct answer is $\dfrac{1}{13^2}$.

Take the reciprocal of the base, with the opposite (positive) exponent:

$13^{-2} = \dfrac{1}{13^2}$

22.

The correct answer is $\dfrac{1}{e^3}$.

Take the reciprocal of the base, with the opposite exponent:

$e^{-3} = \dfrac{1}{e^3}$

23. The correct answer is $\frac{1}{u^7}$.

Take the reciprocal of the base, with a positive exponent:

$$u^{-7} = \frac{1}{u^7}$$

24. The correct answer is $\frac{1}{s^6}$.

Take the reciprocal of the base, with a positive exponent:

$$s^{-6} = \frac{1}{s^6}$$

25. The correct answer is $\frac{1}{p^z}$.

Take the reciprocal of the base, with a positive exponent:

$$p^{-z} = \frac{1}{p^z}$$

Chapter 5

Roots

In this chapter, we'll review the following concepts:

What is a root?
Square roots and cube roots
Operations with roots

What is a root?

A **root** of a number is another number that can be multiplied by itself to produce the original number.

Say we start with an original number, 9.

What number, when multiplied by itself, equals 9?

The answer is 3. In this case, 3 is a root of 9.

Roots are indicated using a symbol called a **radical**:

$$\sqrt{x}$$

The quantity under the radical sign is called the **radicand**. In the example above, x is the radicand.

egghead's Guide to Algebra

Square roots and cube roots

The most common roots we see in algebra are square roots and cube roots.

In this example, 2 is the square root of 4.

Cube roots are multiplied three times to produce another number:

$$3 \times 3 \times 3 = 27$$

Here, the number 3 is the cube root of 27.

There can also be fourth roots, fifth roots, and so on:

$$10 \times 10 \times 10 \times 10 = 10,000$$

The number 10 is the fourth root of 10,000.

When we see a radical sign alone, this indicates the square root of the radicand:

To indicate the cube root or another root, we insert a small number to the upper left of the radical sign:

$$\sqrt[3]{7}$$

This symbol indicates the cube root of 7.

Principal square roots

A root can be a positive or negative number. With square roots, the positive root is known as the **principal square root**. When we see a radical sign by itself, this is taken to indicate the principal square root only.

$$\sqrt{16}$$

The radical shown indicates the positive square root of 16, or 4.

Practice Questions—Square roots
and cube roots

Directions: Simplify the radical expressions shown. You will find the Practice Question Solutions on page 102.

1. $\sqrt{64}$

2. $\sqrt{81}$

3. $\sqrt{225}$

4. $\sqrt[3]{125}$

5. $\sqrt[3]{64}$

6. $\sqrt{x^2}$

7. $\sqrt{r^2}$

8. $\sqrt{y^2}$

9. $\sqrt[3]{a^3}$

10. $\sqrt[3]{z^3}$

Operations with roots

You can calculate the square roots of numbers that aren't perfect squares, but we usually use a calculator to do this. When performing operations with roots, if we have a number that isn't a perfect square or perfect cube, we generally leave that number under the radical sign.

In the last chapter, we learned that operations with exponents must follow certain rules.

The same is true of operations with roots.

Here are the most important rules to know.

For all positive numbers, two numbers that are multiplied under separate radicals can also be combined under one radical:

$$\sqrt{a} \times \sqrt{b} = \sqrt{a \times b}$$

This rule also applies to division:

$$\frac{\sqrt{a}}{\sqrt{b}} = \sqrt{\frac{a}{b}}$$

This doesn't work for addition or subtraction:

$$\sqrt{a} + \sqrt{b} \neq \sqrt{a + b}$$

When you have two numbers added or subtracted under separate radicals, you can't put them both under the same radical sign. Combining only works for multiplication or division. It also works just for numbers greater than one—not negative numbers.

We can use this property to simplify a radical expression.

If you have an expression with numbers multiplied under the same radical, you can separate the factors into individual radicals, as shown:

$$\sqrt{a^2 \times y} = \sqrt{a^2} \times \sqrt{y}$$

The perfect square can then be removed from under the radical sign:

$$\sqrt{a^2 \times y} = \sqrt{a^2} \times \sqrt{y}$$
$$= a\sqrt{y}$$

Examples

Here are some examples of this property at work.

$$\text{Simplify the radical expression } \sqrt{100z}.$$

In this example, we have the number 100 multiplied by the value *z* under the same radical. The number 100 is a perfect square: it breaks down to 10 × 10.

$$\sqrt{100z} = \sqrt{10^2 \times z}$$

The perfect square can now be taken out from under the radical sign:

$$\sqrt{100z} = \sqrt{10^2 \times z}$$
$$= 10\sqrt{z}$$

The variable *z* remains under the radical sign. This expression is in its simplest form.

For another example, let's simplify the radical expression $\sqrt{144e}$.

The number 144 is a perfect square: it is the product of 12 × 12. The 12 can be taken out from under the radical sign. The expression simplifies to $12\sqrt{e}$.

For one last example, we'll use what we've learned to solve an equation:

$$x^2 - 4 = 5$$

This type of equation contains a variable to the second power and is known as a **quadratic equation**. Simplify the equation by moving the 4 to the right hand side:

$$x^2 - 4 = 5$$
$$x^2 - 4 + 4 = 5 + 4$$
$$x^2 = 9$$

Next, isolate the variable. The variable x here is squared. To perform the opposite of that operation, we would take the square root. We must perform this operation on both sides of the equal sign:

$$x^2 = 9$$
$$\sqrt{x^2} = \sqrt{9}$$

On the left-hand side of the equation, the square root of x^2 is x. On the right side of the equation, what number multiplied by itself produces 9? The answer is 3 or –3.

$$\sqrt{x^2} = \sqrt{9}$$
$$\sqrt{x \times x} = \sqrt{9}$$
$$x = \pm\sqrt{9}$$
$$x = \pm 3$$

We use the $\pm$ sign to indicate that the square root of 9 can be positive or negative. When we are simplifying a radical, as we saw above, we give only the principal square root or positive value. When solving an equation with radicals, however, we must show all possible values for the missing number, both positive and negative.

Practice Questions—Operations with roots

Directions: Perform the operations shown. You will find the Practice Question Solutions on page 103.

11. Simplify the radical expression:

$$\sqrt{16a}$$

12. Simplify the radical expression:

$$\sqrt{25y^2}$$

13. Simplify the radical expression:

$$\sqrt[3]{27n}$$

14. Simplify the radical expression:

$$\sqrt[3]{8r^3}$$

15. Simplify the radical expression:

$$\sqrt{48p}$$

16. Simplify the radical expression:

$$\sqrt{60s}$$

17. What are the roots of the equation $n^2 - 2 = 0$?

18. What are the roots of the equation $x^2 - 5 = 1$?

19. Find the root of the equation $a^3 - 7 = 1$.

20. Find the roots of the equation $3y^2 + 4 = 2y^2 + 9$.

Now you've learned the basics of working with radicals and roots. Reinforce your skills with the chapter review.

Chapter Review

Directions: Perform the operations shown. Solutions can be found on page 105.

1. Simplify the radical expression shown.

   $\sqrt{400}$

2. Simplify the radical expression shown.

   $\sqrt{269}$

3. Simplify the radical expression shown.

   $\sqrt{q^2}$

4. Simplify the radical expression shown.

   $\sqrt[3]{216}$

5. Simplify the radical expression shown.

   $\sqrt[3]{t^3}$

6. Write the radical expression $\sqrt{75r}$ in its simplest form.

7. Write the radical expression $\sqrt{44d}$ in its simplest form.

8. Write the radical expressions $\sqrt{256x^2}$ in its simplest form.

9. What is the simplest form of the expression $\sqrt[3]{729m}$?

10. What is the simplest form of the expression $\sqrt{250g}$?

11. Find the roots of the following equation:

    $z^2 + 3 = 16$

12. Find the roots of the following equation:

    $a^2 - 25 = 16$

13. Find the roots of the following equation:

    $n^2 - 34 = 19$

14. Find the roots of the following equation:

    $h^2 - 17 = 12$

15. Find the root of the following equation:

$$p^3 + 7 = 34$$

16. What are the roots of the equation $3e^2 - 4 = 5e^2 - 8$?

17. What are the roots of the equation $12k^2 + 11 = 34k^2 - 11$?

18. Find the roots of the equation $40t^2 + 13 = -9t^2 + 160$.

19. What are the roots of the equation $34a^2 + 7 = 3a^2 + 69$?

20. Find the roots of the equation $2v^2 + 17 = v^2 + 81$.

21. What are the roots of the equation $25 + 8d^2 = 4d^2 + 2d^2 + 145$?

22. What are the roots of the equation $42 + p^2 = 537 - 3p^2 - p^2$?

23. Find the roots of the equation $6m^2 - 73 = 3m^2 - 17 + 2m^2$.

24. What are the roots of the equation $\frac{c^2}{2} = 12$?

25. Find the roots of the equation $\frac{s^2}{4} = 20$.

Practice Question Solutions

Square roots and cube roots

1. The correct answer is 8.

   Here we are simplifying a radical. So, we give the positive square root only.
   $$\sqrt{64} = \sqrt{8 \times 8}$$
   $$= 8$$

2. The correct answer is 9.
   $$\sqrt{81} = \sqrt{9 \times 9}$$
   $$= 9$$

3. The correct answer is 15.
   $$\sqrt{225} = \sqrt{15 \times 15}$$
   $$= 15$$

4.

   The correct answer is 5.

   $$\sqrt[3]{125} = \sqrt[3]{5 \times 5 \times 5}$$
   $$= 5$$

5. The correct answer is 4.
   $$\sqrt[3]{64} = \sqrt[3]{4 \times 4 \times 4}$$
   $$= 4$$

6. The correct answer is x.
   $$\sqrt{x^2} = \sqrt{x \times x}$$
   $$= x$$

7. The correct answer is r.
   $$\sqrt{r^2} = \sqrt{r \times r}$$
   $$= r$$

8. The correct answer is y.
   $$\sqrt{y^2} = \sqrt{y \times y}$$
   $$= y$$

9. The correct answer is a.
   $$\sqrt[3]{a^3} = \sqrt[3]{a \times a \times a}$$
   $$= a$$

10. The correct answer is z.

$$\sqrt[3]{z^3} = \sqrt[3]{z \times z \times z}$$
$$= z$$

Operations with roots

11. The correct answer is $4\sqrt{a}$.

$$\sqrt{16a} = \sqrt{4^2 \times a}$$
$$= \sqrt{4^2} \times \sqrt{a}$$
$$= 4\sqrt{a}$$

12. The correct answer is $5y$.

$$\sqrt{25y^2} = \sqrt{5^2 \times y^2}$$
$$= \sqrt{5^2} \times \sqrt{y^2}$$
$$= 5y$$

13. The correct answer is $3\sqrt[3]{n}$.

$$\sqrt[3]{27n} = \sqrt[3]{3^3 \times n}$$
$$= \sqrt[3]{3^3} \times \sqrt[3]{n}$$
$$= 3\sqrt[3]{n}$$

14. The correct answer is $2r$.

$$\sqrt[3]{8r^3} = \sqrt[3]{2^3 \times r^3}$$
$$= \sqrt[3]{2^3} \times \sqrt[3]{r^3}$$
$$= 2r$$

15. The correct answer is $4\sqrt{3p}$.

$$\sqrt{48p} = \sqrt{16 \times 3 \times p}$$
$$= \sqrt{4^2 \times 3 \times p}$$
$$= \sqrt{4^2} \times \sqrt{3} \times \sqrt{p}$$
$$= 4\sqrt{3p}$$

16. The correct answer is $2\sqrt{15s}$.

$$\sqrt{60s} = \sqrt{4 \times 15 \times s}$$
$$= \sqrt{2^2 \times 15 \times s}$$
$$= \sqrt{2^2} \times \sqrt{15} \times \sqrt{s}$$
$$= 2\sqrt{15s}$$

17. The roots of the equation are $+\sqrt{2}$ and $-\sqrt{2}$.

First, simplify the equation. Get all variables on one side of the equation:

$$n^2 - 2 = 0$$
$$n^2 - 2 + 2 = 0 + 2$$
$$n^2 = 0 + 2$$
$$n^2 = 2$$

Next, isolate the variable. The variable n is squared, so to perform the opposite of that operation, we take the square root:

$$n^2 = 2$$
$$\sqrt{n^2} = \sqrt{2}$$

This lets us isolate the n on the left side of the equation:

$$\sqrt{n^2} = \sqrt{2}$$
$$\sqrt{n \times n} = \sqrt{2}$$
$$n = \pm\sqrt{2}$$

To isolate squares, take the square roots of both sides.

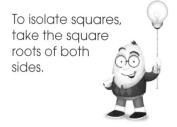

Since the value of n can be positive or negative, we must include the $\pm$ sign. The roots are $+\sqrt{2}$ and $-\sqrt{2}$.

18. The roots of the equation are $+\sqrt{6}$ and $-\sqrt{6}$.

Simplify the equation. Get all variables on one side:

$$x^2 - 5 = 1$$
$$x^2 - 5 + 5 = 1 + 5$$
$$x^2 = 1 + 5$$
$$x^2 = 6$$

Isolate the variable by taking the square root of both sides:

$$x^2 = 6$$
$$\sqrt{x^2} = \sqrt{6}$$

Continue with the solution, including the $\pm$ sign:

$$\sqrt{x^2} = \sqrt{6}$$
$$\sqrt{x \times x} = \sqrt{6}$$
$$x = \pm\sqrt{6}$$

The roots of the equation are $+\sqrt{6}$ and $-\sqrt{6}$.

19. The root of the equation is 2.

Simplify the equation. Add 7 to both sides:

$$a^3 - 7 = 1$$
$$a^3 - 7 + 7 = 1 + 7$$
$$a^3 = 1 + 7$$
$$a^3 = 8$$

Isolate the variable, a, by performing the opposite operation. In this equation, the variable a is cubed. So, we take the cube roots of both sides:

$$a^3 = 8$$
$$\sqrt[3]{a^3} = \sqrt[3]{8}$$
$$a = \sqrt[3]{2^3}$$
$$a = 2$$

The root of the equation is 2.

20. The roots of the equation are $+\sqrt{5}$ and $-\sqrt{5}$.

Simplify the equation. There is no distribution, but like terms can be combined:

$$3y^2 + 4 = 2y^2 + 9$$
$$3y^2 - 2y^2 + 4 = 2y^2 - 2y^2 + 9$$
$$y^2 + 4 = 2y^2 - 2y^2 + 9$$
$$y^2 + 4 = 9$$
$$y^2 + 4 - 4 = 9 - 4$$
$$y^2 = 9 - 4$$
$$y^2 = 5$$

The y variables are now on the left side of the equation, so we can take the square root of both sides:

$$y^2 = 5$$
$$\sqrt{y^2} = \sqrt{5}$$
$$\sqrt{y \times y} = \sqrt{5}$$
$$y = \pm\sqrt{5}$$

Chapter Review Solutions

1.

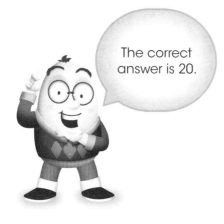

The correct answer is 20.

$$\sqrt{400} = \sqrt{20 \times 20}$$
$$= 20$$

2. The correct answer is 17.
$$\sqrt{289} = \sqrt{17 \times 17}$$
$$= 17$$

3. The correct answer is q.
$$\sqrt{q^2} = \sqrt{q \times q}$$
$$= q$$

4. The correct answer is 6.
$$\sqrt[3]{216} = \sqrt{6 \times 6 \times 6}$$
$$= 6$$

5. The correct answer is t.
$$\sqrt[3]{t^3} = \sqrt{t \times t \times t}$$
$$= t$$

6. The correct answer is $5\sqrt{3r}$.
$$\sqrt{75r} = \sqrt{25 \times 3 \times r}$$
$$= \sqrt{5^2 \times 3 \times r}$$
$$= \sqrt{5^2} \times \sqrt{3} \times \sqrt{r}$$
$$= 5\sqrt{3r}$$

7. The correct answer is $2\sqrt{11d}$.
$$\sqrt{44d} = \sqrt{4 \times 11 \times d}$$
$$= \sqrt{2^2 \times 11 \times d}$$
$$= \sqrt{2^2} \times \sqrt{11} \times \sqrt{d}$$
$$= 2 \times \sqrt{11} \times \sqrt{d}$$
$$= 2\sqrt{11d}$$

8. The correct answer is $16x$.
$$\sqrt{256x^2} = \sqrt{16^2 \times x^2}$$
$$= \sqrt{16^2} \times \sqrt{x^2}$$
$$= 16x$$

9. The correct answer is $9\sqrt[3]{m}$.
$$\sqrt[3]{729m} = \sqrt[3]{9^3 \times m}$$
$$= \sqrt[3]{9} \times \sqrt[3]{m}$$
$$= 9\sqrt[3]{m}$$

10. The correct answer is $5\sqrt{10g}$.

$$\sqrt{250g} = \sqrt{25 \times 10 \times g}$$
$$= \sqrt{5^2 \times 10 \times g}$$
$$= \sqrt{5^2} \times \sqrt{10} \times \sqrt{g}$$
$$= 5\sqrt{10g}$$

11.

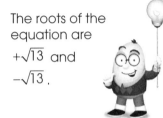

The roots of the equation are $+\sqrt{13}$ and $-\sqrt{13}$.

First, simplify the equation by moving all variables to one side of the equation:

$$z^2 + 3 = 16$$
$$z^2 + 3 - 3 = 16 - 3$$
$$z^2 = 16 - 3$$
$$z^2 = 13$$

Next, isolate the variable. The variable z is squared, so to isolate it, we take the square roots of both sides:

$$z^2 = 13$$
$$\sqrt{z^2} = \sqrt{13}$$

Continue with the solution until z is on the left side of the equation alone. The value of z can be positive or negative, so we include the $\pm$ sign:

$$\sqrt{z^2} = \sqrt{13}$$
$$\sqrt{z \times z} = \sqrt{13}$$
$$z = \pm\sqrt{13}$$

The roots are $+\sqrt{13}$ and $-\sqrt{13}$.

12. The roots of the equation are $+\sqrt{41}$ and $-\sqrt{41}$.

Get all variables on one side of the equation:

$$a^2 - 25 = 16$$
$$a^2 - 25 + 25 = 16 + 25$$
$$a^2 = 16 + 25$$
$$a^2 = 41$$

Isolate the a by taking the square roots of both sides:

$$a^2 = 41$$
$$\sqrt{a^2} = \sqrt{41}$$

There are two possible variables for a, so include the $\pm$ sign:

$$\sqrt{a^2} = \sqrt{41}$$
$$\sqrt{a \times a} = \sqrt{41}$$
$$a = \pm\sqrt{41}$$

The roots are $+\sqrt{41}$ and $-\sqrt{41}$.

13. The roots of the equation are $+\sqrt{53}$ and $-\sqrt{53}$.

Move all variables to the left side of the equation:

$$n^2 - 34 = 19$$
$$n^2 - 34 + 34 = 19 + 34$$
$$n^2 = 19 + 34$$
$$n^2 = 53$$

Take the square roots of both sides:

$$n^2 = 53$$
$$\sqrt{n^2} = \sqrt{53}$$

Continue with the solution, including the ± sign:

$$\sqrt{n^2} = \sqrt{53}$$
$$\sqrt{n \times n} = \sqrt{53}$$
$$n = \pm\sqrt{53}$$

14. The roots of the equation are $+\sqrt{29}$ and $-\sqrt{29}$.

Simplify the equation.

$$h^2 - 17 = 12$$
$$h^2 - 17 + 17 = 12 + 17$$
$$h^2 = 12 + 17$$
$$h^2 = 29$$

Next, take the square roots of both sides:

$$h^2 = 29$$
$$\sqrt{h^2} = \sqrt{29}$$

The value of h can be either positive or negative:

$$\sqrt{h^2} = \sqrt{29}$$
$$\sqrt{h \times h} = \sqrt{29}$$
$$h = \pm\sqrt{29}$$

15. The root of the equation is 3.

Simplify the equation. Subtract 7 from both sides:

$$p^3 + 7 = 34$$
$$p^3 + 7 - 7 = 34 - 7$$
$$p^3 = 34 - 7$$
$$p^3 = 27$$

Isolate the variable by taking the cube roots of both sides:

$$p^3 = 27$$
$$\sqrt[3]{p^3} = \sqrt[3]{27}$$

The number 27 is a perfect cube, so the value of p is an integer:

$$\sqrt[3]{p^3} = \sqrt[3]{27}$$
$$\sqrt[3]{p^3} = \sqrt[3]{3^3}$$
$$p = 3$$

The root of the equation is 3.

16. The roots of the equation are $+\sqrt{2}$ and $-\sqrt{2}$.

First, bring all the variables to the left side of the equation:

$$3e^2 - 4 = 5e^2 - 8$$
$$3e^2 - 5e^2 - 4 = 5e^2 - 5e^2 - 8$$
$$-2e^2 - 4 = 5e^2 - 5e^2 - 8$$
$$-2e^2 - 4 = -8$$

Next, bring all the constants to the right side of the equation:

$$-2e^2 - 4 = -8$$
$$-2e^2 - 4 + 4 = -8 + 4$$
$$-2e^2 = -8 + 4$$
$$-2e^2 = -4$$

Divide both sides of the equation by –2:

$$-2e^2 = -4$$

$$\frac{-2e^2}{-2} = \frac{-4}{-2}$$

$$e^2 = \frac{-4}{-2}$$

$$e^2 = 2$$

Take the square root of both sides:

$$e^2 = 2$$

$$\sqrt{e^2} = \sqrt{2}$$

$$\sqrt{e \times e} = \sqrt{2}$$

$$e = \pm\sqrt{2}$$

The roots of the equation are $+\sqrt{2}$ and $-\sqrt{2}$.

17. The roots of the equation are +1 and –1.

Move the variables to the left side of the equation:

$$12k^2 + 11 = 34k^2 - 11$$

$$-12k^2 - 34k^2 + 11 = 34k^2 - 34k^2 - 11$$

$$-22k^2 + 11 = 34k^2 - 34k^2 - 11$$

$$-22k^2 + 11 = -11$$

Move the constants to the right side:

$$-22k^2 + 11 = -11$$

$$-22k^2 + 11 - 11 = -11 - 11$$

$$-22k^2 = -11 - 11$$

$$-22k^2 = -22$$

Divide both sides of the equation by –22:

$$-22k^2 = -22$$

$$\frac{-22k^2}{-22} = \frac{-22}{-22}$$

$$k^2 = \frac{-22}{-22}$$

$$k^2 = 1$$

Take the square root of both sides:

$$k^2 = 1$$

$$\sqrt{k^2} = \sqrt{1}$$

$$\sqrt{k \times k} = \sqrt{1}$$

$$k = \pm 1$$

The number 1 is a perfect square, so the roots of this equation are integers.

18. The roots of the equation are $+\sqrt{3}$ and $-\sqrt{3}$.

Simplify the equation:

$$40t^2 + 13 = -9t^2 + 160$$

$$40t^2 + 9t^2 + 13 = -9t^2 + 9t^2 + 160$$

$$49t^2 + 13 = -9t^2 + 9t^2 + 160$$

$$49t^2 + 13 = 160$$

$$49t^2 + 13 - 13 = 160 - 13$$

$$49t^2 = 160 - 13$$

$$49t^2 = 147$$

Divide both sides of the equation by 49:

$$49t^2 = 147$$

$$\frac{49t^2}{49} = \frac{147}{49}$$

$$t^2 = \frac{147}{49}$$

$$t^2 = 3$$

Take the square root of both sides:

$$t^2 = 3$$

$$\sqrt{t^2} = \sqrt{3}$$

$$\sqrt{t \times t} = \sqrt{3}$$

$$t = \pm\sqrt{3}$$

19. The roots of the equation are $+\sqrt{2}$ and $-\sqrt{2}$.

Simplify the equation:

$$34a^2 + 7 = 3a^2 + 69$$
$$34a^2 - 3a^2 + 7 = 3a^2 - 3a^2 + 69$$
$$31a^2 + 7 = 3a^2 - 3a^2 + 69$$
$$31a^2 + 7 = 69$$
$$31a^2 + 7 - 7 = 69 - 7$$
$$31a^2 = 69 - 7$$
$$31a^2 = 62$$

Divide both sides of the equation by 31:

$$31a^2 = 62$$
$$\frac{31a^2}{31} = \frac{62}{31}$$
$$a^2 = \frac{62}{31}$$
$$a^2 = 2$$

Take the square root of both sides:

$$a^2 = 2$$
$$\sqrt{a^2} = \sqrt{2}$$
$$\sqrt{a \times a} = \sqrt{2}$$
$$a = \pm\sqrt{2}$$

20. The roots of the equation are +8 and –8.

Simplify the equation:

$$2v^2 + 17 = v^2 + 81$$
$$2v^2 - v^2 + 17 = v^2 - v^2 + 81$$
$$v^2 + 17 = v^2 - v^2 + 81$$
$$v^2 + 17 = 81$$
$$v^2 + 17 - 17 = 81 - 17$$
$$v^2 = 81 - 17$$
$$v^2 = 64$$

Take the square root of both sides:

$$v^2 = 64$$
$$\sqrt{v^2} = \sqrt{64}$$
$$\sqrt{v \times v} = \sqrt{64}$$
$$v = \pm\sqrt{64}$$
$$v = \pm 8$$

21. The roots of the equation are $+2\sqrt{15}$ and $-2\sqrt{15}$.

Simplify:

$$25 + 8d^2 = 4d^2 + 2d^2 + 145$$
$$25 + 8d^2 - 4d^2 - 2d^2 = 4d^2 - 4d^2 + 2d^2 - 2d^2 + 145$$
$$25 + 2d^2 = 4d^2 - 4d^2 + 2d^2 - 2d^2 + 145$$
$$25 + 2d^2 = 145$$
$$25 - 25 + 2d^2 = 145 - 25$$
$$2d^2 = 145 - 25$$
$$2d^2 = 120$$

Divide both sides of the equation by 2:

$$2d^2 = 120$$
$$\frac{2d^2}{2} = \frac{120}{2}$$
$$d^2 = \frac{120}{2}$$
$$d^2 = 60$$

Take the square root of both sides:

$$d^2 = 60$$
$$\sqrt{d^2} = \sqrt{60}$$
$$\sqrt{d \times d} = \sqrt{60}$$
$$d = \pm\sqrt{60}$$

The number 60 can be further factored into 4 × 15, so we continue with this step:

$$d = \pm\sqrt{60}$$
$$= \pm\sqrt{4 \times 15}$$
$$= \pm\sqrt{2^2 \times 15}$$
$$= \pm\sqrt{2^2} \times \sqrt{15}$$
$$= \pm 2\sqrt{15}$$

The roots of the equation are $+2\sqrt{15}$ and $-2\sqrt{15}$.

22.

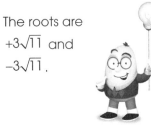

The roots are $+3\sqrt{11}$ and $-3\sqrt{11}$.

Simplify:

$$42 + p^2 = 537 - 3p^2 - p^2$$
$$42 + p^2 + 3p^2 + p^2 = 537 - 3p^2 - p^2 + 3p^2 + p^2$$
$$42 + 5p^2 = 537 - 3p^2 - p^2 + 3p^2 + p^2$$
$$42 + 5p^2 = 537$$
$$42 - 42 + 5p^2 = 537 - 42$$
$$5p^2 = 537 - 42$$
$$5p^2 = 495$$

Divide both sides of the equation by 5:

$$5p^2 = 495$$
$$\frac{5p^2}{5} = \frac{495}{5}$$
$$p^2 = \frac{495}{5}$$
$$p^2 = 99$$

Take the square root of both sides:

$$p^2 = 99$$
$$\sqrt{p^2} = \sqrt{99}$$
$$\sqrt{p \times p} = \sqrt{99}$$
$$p = \pm\sqrt{99}$$

We can factor the number 99 into a perfect square, 9, times 11:

$$p = \pm\sqrt{99}$$
$$= \pm\sqrt{9 \times 11}$$
$$= \pm\sqrt{3^2 \times 11}$$
$$= \pm\sqrt{3^2} \times \sqrt{11}$$
$$= \pm 3\sqrt{11}$$

23. The roots of the equation are $+2\sqrt{14}$ and $-2\sqrt{14}$.

First, simplify the equation:

$$6m^2 - 73 = 3m^2 - 17 + 2m^2$$
$$6m^2 - 3m^2 - 2m^2 - 73 = 3m^2 - 3m^2 - 17 + 2m^2 - 2m^2$$
$$m^2 - 73 = 3m^2 - 3m^2 - 17 + 2m^2 - 2m^2$$
$$m^2 - 73 = -17$$
$$m^2 - 73 + 73 = -17 + 73$$
$$m^2 = -17 + 73$$
$$m^2 = 56$$

Then take the square root of both sides:

$$m^2 = 56$$
$$\sqrt{m^2} = \sqrt{56}$$
$$\sqrt{m \times m} = \sqrt{56}$$
$$m = \pm\sqrt{56}$$
$$= \pm\sqrt{4 \times 14}$$
$$= \pm\sqrt{2^2 \times 14}$$
$$= \pm\sqrt{2^2} \times \sqrt{14}$$
$$= \pm 2\sqrt{14}$$

There are no perfect squares in the number 14, so this is the final answer.

24. The roots of the equation are $+2\sqrt{6}$ and $-2\sqrt{6}$.

Multiply both sides of the equation by 2 to get rid of the 2 in the denominator:

$$\frac{c^2}{2} = 12$$
$$2 \times \left(\frac{c^2}{2}\right) = 12 \times 2$$
$$c^2 = 12 \times 2$$
$$c^2 = 24$$

Now, take the square root of both sides:

$$c^2 = 24$$
$$\sqrt{c^2} = \sqrt{24}$$
$$\sqrt{c \times c} = \sqrt{24}$$
$$c = \pm\sqrt{24}$$

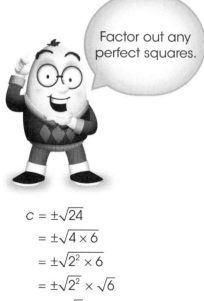

Factor out any perfect squares.

$$c = \pm\sqrt{24}$$
$$= \pm\sqrt{4 \times 6}$$
$$= \pm\sqrt{2^2 \times 6}$$
$$= \pm\sqrt{2^2} \times \sqrt{6}$$
$$= \pm 2\sqrt{6}$$

25. The roots of the equation are $+4\sqrt{5}$ and $-4\sqrt{5}$.

Multiply both sides of the equation by 4:

$$\frac{s^2}{4} = 20$$
$$4 \times \left(\frac{s^2}{4}\right) = 20 \times 4$$
$$s^2 = 20 \times 4$$
$$s^2 = 80$$

Take the square root of both sides:

$$s^2 = 80$$
$$\sqrt{s^2} = \sqrt{80}$$
$$\sqrt{s \times s} = \sqrt{80}$$
$$s = \pm\sqrt{80}$$
$$s = \pm\sqrt{16 \times 5}$$
$$= \pm\sqrt{4^2 \times 5}$$
$$= \pm\sqrt{4^2} \times \sqrt{5}$$
$$= \pm 4\sqrt{5}$$

The roots are $+4\sqrt{5}$ and $-4\sqrt{5}$.

Chapter 6

Polynomials

In this chapter, we'll review the following concepts:

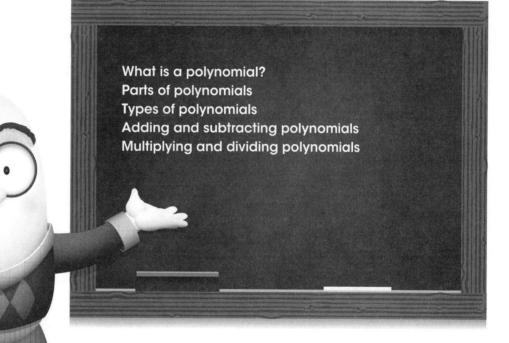

What is a polynomial?
Parts of polynomials
Types of polynomials
Adding and subtracting polynomials
Multiplying and dividing polynomials

What is a polynomial?

A **polynomial** is the sum of expressions containing variables and exponents.

$$3x^2 + x + 25$$

The exponents in a polynomial expression can only be positive integers or zero. There can be no negative exponents or exponents made up of fractions.

NOT polynomials:

$$4y^{-2} + 3y + 7$$

$$5a^{\frac{1}{3}} + 2a^2 + 6a + 3$$

We haven't worked with fractional exponents yet. But they cannot be contained in polynomials. Polynomials also follow some other rules:

- There can be no radicals with variables; and
- There can be no fractions with variables in the denominator.

NOT polynomials:

$$2n^3 + 6\sqrt{n} + 7$$

$$15r^2 + \frac{19}{r} + 2$$

Parts of polynomials

Polynomials consist of terms and the operators that connect them.

The **terms** of a polynomial are the parts that are added:

$$6x^3 + 2x^2 + 7x + 5$$

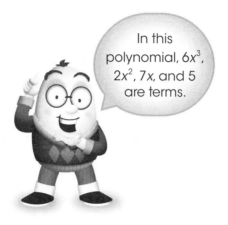

In this polynomial, $6x^3$, $2x^2$, $7x$, and 5 are terms.

The **operators** of a polynomial are the symbols for addition or subtraction:

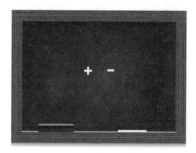

Subtraction is the same as adding a negative. When we write the polynomial $7x - 3$, we're really writing:

$$7x + (-3)$$

Polynomials can also be described in terms of **degree**. The degree of a polynomial containing one variable is the power to which the variable is raised.

This polynomial is a second-degree polynomial. The term $8n^2$ has an exponent of 2.

If a polynomial has more than one variable raised to an exponent, the polynomial has the degree of the largest exponent:

$$17y^3 + 3y^2 + 2y$$

This polynomial is a third-degree polynomial. Its largest exponent is 3.

If a polynomial has a term with two exponents, we add the exponents to determine the degree:

$$4x^3y^2$$

This polynomial is a fifth-degree polynomial. The sum of its exponents, $3 + 2$, is 5.

The terms of polynomials are usually ordered in descending order from left to right, beginning with the highest degree. This is the standard form of a polynomial expression.

$$6n^3 + 2n^2 + 5n + 1$$

Types of polynomials

Certain polynomials have special names. Here are the names for the most common polynomials you're likely to see.

A **monomial** is a product of numbers or variables.

$$4x^3y^2$$

The example $4x^3y^2$ is a monomial. It contains just one term.

A **binomial** is the sum of two monomials. It contains two unlike terms.

$$5c^2 + 8c$$

In the example shown, $5c^2$ and $8c$ are unlike terms, because they have different exponents.

A **trinomial** is the sum of three monomials. It contains three unlike terms:

$$9r^2 + 12r + 13$$

The term **polynomial** is used as a general name for these types of expressions, regardless of the number of terms. It is also used to refer to the sum of more than three monomials.

Monomials, binomials, and trinomials are all polynomials.

Polynomials can also be classified by different names, based on their degree. First-degree polynomials are called **linear** polynomials. Second-degree polynomials are called **quadratics**, and polynomials in the third degree are known as **cubics**.

Here's a chart to help you remember.

Name	What it Means
By number of terms	
Monomial	Single term
Binomial	Sum of two monomials
Trinomial	Sum of three monomials
Polynomial	Sum of four or more monomials; General name for all
By degree	
Linear	First-degree polynomial
Quadratic	Second-degree polynomial
Cubic	Third-degree polynomial

Adding and subtracting polynomials

When we add and subtract polynomials, we combine like terms. You should be familiar with this process, as we practiced it in earlier chapters.

Let's review the rules.

Like terms must have the same variables, and they must be raised to the same power. Constants are like terms, so they can also be combined.

The following are examples of expressions containing like terms:

$$5a^2 + 6a^2$$

$$7x^3 + 7x^3$$

$$2m^2n^2 + 10m^2n^2$$

To combine these, we add the coefficients:

$$5a^2 + 6a^2 = 11a^2$$

$$7x^3 + 7x^3 = 14x^3$$

$$2m^2n^2 + 10m^2n^2 = 12\ m^2n^2$$

The following expressions do NOT contain like terms:

$$c^2 + c^3$$

$$ab + a^2b^2$$

These terms cannot be combined.

Examples

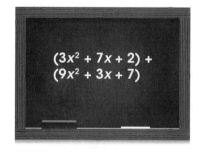

Here are a few examples.

To add these polynomials, first group them by moving like terms next to each other:

$$(3x^2 + 7x + 2) + (9x^2 + 3x + 7)$$

$$3x^2 + 9x^2 + 7x + 3x + 2 + 7$$

Combine all the terms with x^2:

$$3x^2 + 9x^2 + 7x + 3x + 2 + 7$$

$$12x^2 + 7x + 3x + 2 + 7$$

Combine all the terms with x:

$$12x^2 + 7x + 3x + 2 + 7$$

$$12x^2 + 10x + 2 + 7$$

Combine all the constants:

$$12x^2 + 10x + 2 + 7$$

$$12x^2 + 10x + 9$$

The simplified polynomial is $12x^2 + 10x + 9$.

Here's an example using subtraction:

$$(7a^2 + 3a + 4) - (2a^3 + 4a + 2)$$

Group like terms together. When you remove the parentheses, change the signs to show subtraction.

$$(7a^2 + 3a + 4) - (2a^3 + 4a + 2)$$

$$7a^2 + 3a + 4 - 2a^3 - 4a - 2$$

$$-2a^3 + 7a^2 + 3a - 4a + 4 - 2$$

There is only one term with a^3, so this one goes first. There is just one term with a^2, so this is placed second.

Combine all terms with a:

$$-2a^3 + 7a^2 + 3a - 4a + 4 - 2$$

$$-2a^3 + 7a^2 - a + 4 - 2$$

Combine constants:

$$-2a^3 + 7a^2 - a + 2$$

The simplified polynomial is $-2a^3 + 7a^2 - a + 2$.

Practice Questions—Adding and subtracting polynomials

Directions: Simplify the polynomials using addition or subtraction. Practice Question Solutions begin on page 128.

1. $(2x + 17) + (4x^2 + 3)$

2. $(5y^3 + 12y) + (7y^2 + 16)$

3. $(10n^4 + 26n) - (n^3 + 3n^2)$

4. $(c^3 - 6c^2) - (9c^3 - 5)$

5. $(a^2 + 7a^3 + 4) + (6 + 3a^3 + 11a)$

6. $(k^3 + 9k^2 + 10) + (3 + 5k^3 + 20)$

7. $(5r^2 + 7r + 4r^4) + (6r^3 + 3r^4 + 10r)$

8. $(25b^2 + 15b + 6) - (12b^2 - 4b - 8)$

Be careful with your subtraction here!

9. $(5p^2q^2 + 15p + 12pq + 3) - (14q^2 - pq + 6p - 91)$

10. $(2c^2d^2 + 3cd + 9c - 7) - (6c^2 - 10c^2d^2 - 21cd + 7c)$

Multiplying and dividing polynomials

Now that we've seen addition and subtraction, let's move on to multiplication and division. Multiplication and division of polynomials can get a little more complex.

Multiplying by monomials

Remember the rules of working with exponents?

When you multiply exponential expressions with the same base, the exponents are added:

$$3x^2(4x^2 + 5) = 12x^4 + 15x^2$$

Here, for the first two terms, we multiplied the coefficients: 3 × 4. We kept the bases the same, x, and we added the exponents: 2 + 2 = 4.

Don't multiply the exponents!

It's important to remember that exponents get **added** when terms are multiplied. The only time we multiply exponents is when we are raising a power to another power.

Dividing by monomials

When we divide by monomials, we treat the fraction as if it were separate fractions. Create separate fractions for each term in the numerator:

$$\frac{12x^2 + 9x}{3x} = \frac{12x^2}{3x} + \frac{9x}{3x}$$

Perform the division separately for each fraction:

$$\frac{12x^2 + 9x}{3x} = \frac{12x^2}{3x} + \frac{9x}{3x}$$

$$= 4x^{2-1} + \frac{9x}{3x}$$

$$= 4x + \frac{9x}{3x}$$

$$= 4x + 3$$

When dividing exponents with the same base, the exponents are subtracted.

Then add the terms back together.

Multiplying binomials

A special technique is used for multiplying two binomials. This technique is called FOIL.

FOIL stands for **F**irst, **O**uter, **I**nner, **L**ast.

The term FOIL is a way to remember the order of operations when multiplying binomials.

Say we want to multiply the binomials $(n + 3)(n + 2)$.

With FOIL, we multiply the first, outer, inner, and last terms as follows:

FIRST	$(\boldsymbol{n} + 3)(\boldsymbol{n} + 2)$	$n \times n = n^2$
OUTER	$(\boldsymbol{n} + 3)(n + \boldsymbol{2})$	$n \times 2 = 2n$
INNER	$(n + \boldsymbol{3})(\boldsymbol{n} + 2)$	$3 \times n = 3n$
LAST	$(n + \boldsymbol{3})(n + \boldsymbol{2})$	$3 \times 2 = 6$

Then add the products together:

$$n^2 + 2n + 3n + 6$$

Finally, simplify:

$$n^2 + 5n + 6$$

An important note about FOIL!

The FOIL technique can only be used when multiplying two binomials. It will not work when multiplying other types of polynomials.

Multiplying binomials and trinomials

To multiply binomials and trinomials, we use distribution.

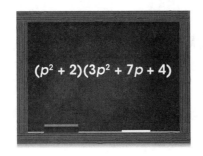

$$(p^2 + 2)(3p^2 + 7p + 4)$$

1 Distribute the p^2 over the three terms in the trinomial:

$$(p^2 \times 3p^2) + (p^2 \times 7p) + (p^2 \times 4)$$

$$(3p^{2+2}) + (7p^{2+1}) + (4p^2)$$

$$3p^4 + 7p^3 + 4p^2$$

2 Then, distribute the 2:

$$(2 \times 3p^2) + (2 \times 7p) + (2 \times 4)$$

$$6p^2 + 14p + 8$$

3 Add the products together:

$$3p^4 + 7p^3 + 4p^2 + 6p^2 + 14p + 8$$

4 Put the terms in degree order:

$$3p^4 + 7p^3 + 4p^2 + 6p^2 + 14p + 8$$

In this case, they already are in degree order, so move on to the next step.

5 Combine like terms:

$$3p^4 + 7p^3 + 10p^2 + 14p + 8$$

We'll work more with dividing polynomials in the next chapter. For now, let's practice what we've learned.

Practice Questions—Multiplying and dividing polynomials

Directions: Simplify the polynomials using multiplication or division. Practice Question Solutions begin on page 130.

11. $7a^3(6a^2 + 2)$

12. $9y(8y^3 + 7y^2 + 6)$

13. $\dfrac{16b^2 + 12b}{4b}$

14. $\dfrac{144n^3 + 24n^2}{12n^2}$

15. $(g + 8)(g + 6)$

16. $(r + 2)(r + 1)$

17. $(u + 9)(u + 13)$

18. $(s^2 + 12s)(2s^2 + 11s + 5)$

19. $(2z^2 + 15)(6z^2 + 31z - 7)$

20. $(t + 4)^3$

Chapter Review

Directions: Simplify the polynomials using addition, subtraction, multiplication, or division. Chapter Review Solutions begin on page 133.

1. $(2g^4 + 16g) + (12g^4 + 3g)$

2. $(9p^6 + 7p^2) + (p^2 + 2)$

3. $(3e^4 - 4d^3) - (2e^7 - 8)$

4. $(5q^2 + 6q^2 + 32) + (q^2 + 23q + 7)$

5. $(3u^4 + 4u^2 + 4u) + (5u^2 + u + 17)$

6. $7x(2x^2 + 9)$

7. $6a^3(4a^3 + 3)$

8. $n(5n^2 + 7n)$

9. $6h^2(2h^3 + 4h)$

10. $3j(14j^3 + 12j^2)$

11. $\dfrac{2y + 6}{2}$

12. $\dfrac{14k^2 + 28k}{7k}$

13. $\dfrac{12b^3 - 9b^2}{3b}$

14. $\dfrac{38r^3 + 19r^2}{19r^2}$

15. $\dfrac{28v^2 + 49}{7}$

16. $(c + 10)(c + 6)$

17. $(w + 5)(w + 4)$

18. $(3d + 2)(d - 3)$

19. $(4s + 6)(9s - 7)$

20. $(2m - 1)(m - 5)$

21. $(21g^3 + 7)(5g^3 + 15g^2 + 6)$

22. $(99r^2 + 11r)(9r^2 + r - 12)$

23. $(7x^2 + 4)(6x^3 - 3x^2 + 2)$

24. $(t^2 - 4t)(7t^2 + 9t + 1)$

25. $(4a^2 - 2)(5a^2 - 26a + 27)$

Practice Question Solutions

Adding and subtracting polynomials

1. The correct answer is $4x^2 + 2x + 20$.

    $$(2x + 17) + (4x^2 + 3) = 2x + 17 + 4x^2 + 3$$
    $$= 4x^2 + 2x + 17 + 3$$
    $$= 4x^2 + 2x + 20$$

2. The correct answer is $5y^3 + 7y^2 + 12y + 16$.

    $$(5y^3 + 12y) + (7y^2 + 16) = 5y^3 + 12y + 7y^2 + 16$$
    $$= 5y^3 + 7y^2 + 12y + 16$$

3. The correct answer is $10n^4 - n^3 - 3n^2 + 26n$.

    $$(10n^4 + 26n) - (n^3 + 3n^2) = 10n^4 + 26n - n^3 - 3n^2$$
    $$= 10n^4 - n^3 - 3n^2 + 26n$$

4.

The correct answer is $-8c^3 - 6c^2 + 5$.

$$(c^3 - 6c^2) - (9c^3 - 5) = c^3 - 6c^2 - 9c^3 + 5$$
$$= c^3 - 9c^3 - 6c^2 + 5$$
$$= -8c^3 - 6c^2 + 5$$

5. The correct answer is $10a^3 + a^2 + 11a + 10$.

$$(a^2 + 7a^3 + 4) + (6 + 3a^3 + 11a) = a^2 + 7a^3 + 4 + 6 + 3a^3 + 11a$$
$$= 7a^3 + 3a^3 + a^2 + 11a + 4 + 6$$
$$= 10a^3 + a^2 + 11a + 4 + 6$$
$$= 10a^3 + a^2 + 11a + 10$$

6. The correct answer is $6k^3 + 9k^2 + 33$.

$$(k^3 + 9k^2 + 10) + (3 + 5k^3 + 20) = k^3 + 9k^2 + 10 + 3 + 5k^3 + 20$$
$$= k^3 + 5k^3 + 9k^2 + 10 + 3 + 20$$
$$= 6k^3 + 9k^2 + 10 + 3 + 20$$
$$= 6k^3 + 9k^2 + 33$$

7. The correct answer is $7r^4 + 6r^3 + 5r^2 + 17r$.

$$(5r^2 + 7r + 4r^4) + (6r^3 + 3r^4 + 10r) = 5r^2 + 7r + 4r^4 + 6r^3 + 3r^4 + 10r$$
$$= 4r^4 + 3r^4 + 6r^3 + 5r^2 + 7r + 10r$$
$$= 7r^4 + 6r^3 + 5r^2 + 7r + 10r$$
$$= 7r^4 + 6r^3 + 5r^2 + 17r$$

8. The correct answer is $13b^2 + 19b + 14$.

    In this question, the second trinomial is subtracted from the first. Subtracting a negative is the same as adding a positive. When we remove the parentheses, we must change the minus signs to plus signs for the last two terms, $4b$ and 8.

$$(25b^2 + 15b + 6) - (12b^2 - 4b - 8) = 25b^2 + 15b + 6 - 12b^2 + 4b + 8$$
$$= 25b^2 - 12b^2 + 15b + 4b + 6 + 8$$
$$= 13b^2 + 15b + 4b + 6 + 8$$
$$= 13b^2 + 19b + 6 + 8$$
$$= 13b^2 + 19b + 14$$

    When subtracting negative terms, pay close attention to the plus and minus signs.

9. The correct answer is $5p^2q^2 - 14q^2 + 13pq + 9p + 94$.

$$(5p^2q^2 + 15p + 12pq + 3) - (14q^2 - pq + 6p - 91) = 5p^2q^2 + 15p + 12pq + 3 - 14q^2 + pq - 6p + 91$$
$$= 5p^2q^2 - 14q^2 + 12pq + pq + 15p - 6p + 3 + 91$$
$$= 5p^2q^2 - 14q^2 + 13pq + 15p - 6p + 3 + 91$$
$$= 5p^2q^2 - 14q^2 + 13pq + 9p + 3 + 91$$
$$= 5p^2q^2 - 14q^2 + 13pq + 9p + 94$$

10. The correct answer is $12c^2d^2 - 6c^2 + 24cd + 2c - 7$.

$$(2c^2d^2 + 3cd + 9c - 7) - (6c^2 - 10c^2d^2 - 21cd + 7c) = 2c^2d^2 + 3cd + 9c - 7 - 6c^2 + 10c^2d^2 + 21cd - 7c$$
$$= 2c^2d^2 + 10c^2d^2 - 6c^2 + 3cd + 21cd + 9c - 7c - 7$$
$$= 12c^2d^2 - 6c^2 + 3cd + 21cd + 9c - 7c - 7$$
$$= 12c^2d^2 - 6c^2 + 24cd + 9c - 7c - 7$$
$$= 12c^2d^2 - 6c^2 + 24cd + 2c - 7$$

Multiplying and dividing polynomials

11. The correct answer is $42a^5 + 14a^3$.

Distribute the $7a^3$ over the $6a^2$ and the 2:

$$7a^3\left(6a^2 + 2\right) = \left(7a^3 \times 6a^2\right) + \left(7a^3 \times 2\right)$$
$$= 42a^{3+2} + 14a^3$$
$$= 42a^5 + 14a^3$$

12. The correct answer is $72y^4 + 63y^3 + 54y$.

$$9y\left(8y^3 + 7y^2 + 6\right) = \left(9y \times 8y^3\right) + \left(9y \times 7y^2\right) + \left(9y \times 6\right)$$
$$= 72y^{1+3} + 63y^{1+2} + 54y$$
$$= 72y^4 + 63y^3 + 54y$$

13. The correct answer is $4b + 3$.

$$\frac{16b^2 + 12b}{4b} = \frac{16b^2}{4b} + \frac{12b}{4b}$$
$$= 4b^{2-1} + \frac{12b}{4b}$$
$$= 4b + \frac{12b}{4b}$$
$$= 4b + 3$$

14. The correct answer is $12n + 2$.

$$\frac{144n^3 + 24n^2}{12n^2} = \frac{144n^3}{12n^2} + \frac{24n^2}{12n^2}$$
$$= 12n^{3-2} + \frac{24n^2}{12n^2}$$
$$= 12n + \frac{24n^2}{12n^2}$$
$$= 12n + 2$$

15. The correct answer is $g^2 + 14g + 48$.

$$(g + 8)(g + 6) = (g \times g) + (g \times 6) + (8 \times g) + (8 \times 6)$$
$$= g^2 + 6g + 8g + 48$$
$$= g^2 + 14g + 48$$

16. The correct answer is $r^2 + 3r + 2$.

$$(r + 2)(r + 1) = (r \times r) + (r \times 1) + (2 \times r) + (2 \times 1)$$
$$= r^2 + r + 2r + 2$$
$$= r^2 + 3r + 2$$

17. The correct answer is $u^2 + 22u + 117$.

$$(u + 9)(u + 13) = (u \times u) + (u \times 13) + (9 \times u) + (9 \times 13)$$
$$= u^2 + 13u + 9u + 117$$
$$= u^2 + 22u + 117$$

18. The correct answer is $2s^4 + 35s^3 + 137s^2 + 60s$.

$$(s^2 + 12s)(2s^2 + 11s + 5) = (s^2 \times 2s^2) + (s^2 \times 11s) + (s^2 \times 5) + (12s \times 2s^2) + (12s \times 11s) + (12s \times 5)$$
$$= (2s^{2+2}) + (11s^{2+1}) + (5s^2) + (24s^{1+2}) + (132s^{1+1}) + (60s)$$
$$= 2s^4 + 11s^3 + 5s^2 + 24s^3 + 132s^2 + 60s$$
$$= 2s^4 + 11s^3 + 24s^3 + 5s^2 + 132s^2 + 60s$$
$$= 2s^4 + 35s^3 + 137s^2 + 60s$$

19. The correct answer is $12z^4 + 62z^3 + 76z^2 + 465z - 105$.

$$(2z^2 + 15)(6z^2 + 31z - 7) = (2z^2 \times 6z^2) + (2z^2 \times 31z) + [2z^2 \times (-7)] + (15 \times 6z^2) + (15 \times 31z) + [15 \times (-7)]$$
$$= (12z^{2+2}) + (62z^{2+1}) + (-14z^2) + (90z^2) + (465z) + (-105)$$
$$= 12z^4 + 62z^3 - 14z^2 + 90z^2 + 465z - 105$$
$$= 12z^4 + 62z^3 + 76z^2 + 465z - 105$$

20. The correct answer is $t^3 + 12t^2 + 48t + 64$.

$$(t+4)^3 = (t+4)(t+4)(t+4)$$

Use FOIL to multiply the first two binomials.

$$(t+4)(t+4) = (t \times t) + (t \times 4) + (4 \times t) + (4 \times 4)$$
$$= t^2 + 4t + 4t + 16$$
$$= t^2 + 8t + 16$$

Using FOIL on the first two binomials gives the trinomial $t^2 + 8t + 16$. Now, use distribution:

$$(t+4)(t^2 + 8t + 16) = (t \times t^2) + (t \times 8t) + (t \times 16) + (4 \times t^2) + (4 \times 8t) + (4 \times 16)$$
$$= (t^{1+2}) + (8t^{1+1}) + (16t) + (4t^2) + (32t) + (64)$$
$$= t^3 + 8t^2 + 16t + 4t^2 + 32t + 64$$
$$= t^3 + 8t^2 + 4t^2 + 16t + 32t + 64$$
$$= t^3 + 12t^2 + 48t + 64$$

Chapter Review Solutions

1. The correct answer is $14g^4 + 19g$.

$$\left(2g^4 + 16g\right) + \left(12g^4 + 3g\right) = 2g^4 + 16g + 12g^4 + 3g$$
$$= 2g^4 + 12g^4 + 16g + 3g$$
$$= 14g^4 + 19g$$

2. The correct answer is $9p^6 + 8p^2 + 2$.

$$\left(9p^6 + 7p^2\right) + \left(p^2 + 2\right) = 9p^6 + 7p^2 + p^2 + 2$$
$$= 9p^6 + 8p^2 + 2$$

3. The correct answer is $-2e^7 + 3e^4 - 4d^3 + 8$.

$$\left(3e^4 - 4d^3\right) - \left(2e^7 - 8\right) = 3e^4 - 4d^3 - 2e^7 + 8$$
$$= -2e^7 + 3e^4 - 4d^3 + 8$$

4. The correct answer is $12q^2 + 23q + 39$.

$$\left(5q^2 + 6q^2 + 32\right) + \left(q^2 + 23q + 7\right) = 5q^2 + 6q^2 + 32 + q^2 + 23q + 7$$
$$= 5q^2 + 6q^2 + q^2 + 23q + 32 + 7$$
$$= 12q^2 + 23q + 32 + 7$$
$$= 12q^2 + 23q + 39$$

5. The correct answer is $3u^4 + 9u^2 + 5u + 17$.

$$\left(3u^4 + 4u^2 + 4u\right) + \left(5u^2 + u + 17\right) = 3u^4 + 4u^2 + 4u + 5u^2 + u + 17$$
$$= 3u^4 + 4u^2 + 5u^2 + 4u + u + 17$$
$$= 3u^4 + 9u^2 + 4u + u + 17$$
$$= 3u^4 + 9u^2 + 5u + 17$$

6. The correct answer is $14x^3 + 63x$.

$$7x\left(2x^2 + 9\right) = \left(7x \times 2x^2\right) + \left(7x \times 9\right)$$
$$= \left(14x^{1+2}\right) + \left(63x\right)$$
$$= 14x^3 + 63x$$

7. The correct answer is $24a^6 + 18a^3$.

$$6a^3\left(4a^3 + 3\right) = \left(6a^3 \times 4a^3\right) + \left(6a^3 \times 3\right)$$
$$= \left(24a^{3+3}\right) + \left(18a^3\right)$$
$$= 24a^6 + 18a^3$$

8. The correct answer is $5n^3 + 7n^2$.

$$n\left(5n^2 + 7n\right) = \left(n \times 5n^2\right) + \left(n \times 7n\right)$$
$$= \left(5n^{1+2}\right) + \left(7n^{1+1}\right)$$
$$= 5n^3 + 7n^2$$

9. The correct answer is $12h^5 + 24h^3$.

$$6h^2\left(2h^3 + 4h\right) = \left(6h^2 \times 2h^3\right) + \left(6h^2 \times 4h\right)$$
$$= \left(12h^{2+3}\right) + \left(24h^{2+1}\right)$$
$$= 12h^5 + 24h^3$$

10. The correct answer is $42j^4 + 36j^3$.

$$3j\left(14j^3 + 12j^2\right) = \left(3j \times 14j^3\right) + \left(3j \times 12j^2\right)$$
$$= \left(42j^{1+3}\right) + \left(36j^{1+2}\right)$$
$$= 42j^4 + 36j^3$$

11.

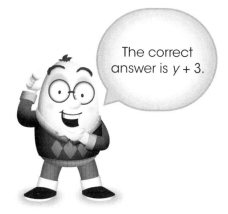

The correct answer is $y + 3$.

$$\frac{2y + 6}{2} = \frac{2y}{2} + \frac{6}{2}$$
$$= y + \frac{6}{2}$$
$$= y + 3$$

12. The correct answer is $2k + 4$.

$$\frac{14k^2 + 28k}{7k} = \frac{14k^2}{7k} + \frac{28k}{7k}$$
$$= \left(2k^{2-1}\right) + 4$$
$$= 2k + 4$$

13. The correct answer is $4b^2 - 3b$.

$$\frac{12b^3 - 9b^2}{3b} = \frac{12b^3}{3b} - \frac{9b^2}{3b}$$
$$= \left(4b^{3-1}\right) - \left(3b^{2-1}\right)$$
$$= 4b^2 - 3b$$

14. The correct answer is $2r + 1$.

$$\frac{38r^3 + 19r^2}{19r^2} = \frac{38r^3}{19r^2} + \frac{19r^2}{19r^2}$$
$$= \left(2r^{3-2}\right) + 1$$
$$= 2r + 1$$

15. The correct answer is $4v^2 + 7$.

$$\frac{28v^2 + 49}{7} = \frac{28v^2}{7} + \frac{49}{7}$$
$$= 4v^2 + \frac{49}{7}$$
$$= 4v^2 + 7$$

16. The correct answer is $c^2 + 16c + 60$.

$$(c + 10)(c + 6) = (c \times c) + (c \times 6) + (10 \times c) + (10 \times 6)$$
$$= c^2 + 6c + 10c + 60$$
$$= c^2 + 16c + 60$$

17. The correct answer is $w^2 + 9w + 20$.

$$(w + 5)(w + 4) = (w \times w) + (w \times 4) + (5 \times w) + (5 \times 4)$$
$$= w^2 + 4w + 5w + 20$$
$$= w^2 + 9w + 20$$

18. The correct answer is $3d^2 - 7d - 6$.

$$(3d + 2)(d - 3) = (3d \times d) + \left[3d \times (-3)\right] + (2 \times d) + \left[2 \times (-3)\right]$$
$$= \left(3d^2\right) + (-9d) + (2d) + (-6)$$
$$= 3d^2 - 9d + 2d - 6$$
$$= 3d^2 - 7d - 6$$

19. The correct answer is $36s^2 + 26s - 42$.

$$(4s + 6)(9s - 7) = (4s \times 9s) + \left[4s \times (-7)\right] + (6 \times 9s) + \left[6 \times (-7)\right]$$
$$= \left(36s^2\right) + (-28s) + (54s) + (-42)$$
$$= 36s^2 - 28s + 54s - 42$$
$$= 36s^2 + 26s - 42$$

20. The correct answer is $2m^2 - 11m + 6$.

$$(2m - 1)(m - 5) = (2m \times m) + \left[2m \times (-5)\right] + \left[(-1) \times m\right] + \left[(-1) \times (-5)\right]$$
$$= \left(2m^2\right) + (-10m) + (-m) + (6)$$
$$= 2m^2 - 10m - m + 6$$
$$= 2m^2 - 11m + 6$$

21. The correct answer is $105g^6 + 315g^5 + 161g^3 + 105g^2 + 42$.

$$\left(21g^3 + 7\right)\left(5g^3 + 15g^2 + 6\right) = \left(21g^3 \times 5g^3\right) + \left(21g^3 \times 15g^2\right) + \left(21g^3 \times 6\right) + \left(7 \times 5g^3\right) + \left(7 \times 15g^2\right) + \left(7 \times 6\right)$$
$$= \left(105g^{3+3}\right) + \left(315g^{3+2}\right) + \left(126g^3\right) + \left(35g^3\right) + \left(105g^2\right) + \left(42\right)$$
$$= 105g^6 + 315g^5 + 126g^3 + 35g^3 + 105g^2 + 42$$
$$= 105g^6 + 315g^5 + 161g^3 + 105g^2 + 42$$

22. The correct answer is $891r^4 + 198r^3 - 1,177r^2 - 132r$.

$$\left(99r^2 + 11r\right)\left(9r^2 + r - 12\right) = \left(99r^2 \times 9r^2\right) + \left(99r^2 \times r\right) + \left[99r^2 \times (-12)\right] + \left(11r \times 9r^2\right) + \left(11r \times r\right) + \left[11r \times (-12)\right]$$
$$= \left(891r^{2+2}\right) + \left(99r^{2+1}\right) + \left(-1,188r^2\right) + \left(99r^{1+2}\right) + \left(11r^{1+1}\right) + \left(-132r\right)$$
$$= 891r^4 + 99r^3 - 1,188r^2 + 99r^3 + 11r^2 - 132r$$
$$= 891r^4 + 99r^3 + 99r^3 - 1,188r^2 + 11r^2 - 132r$$
$$= 891r^4 + 198r^3 - 1,188r^2 + 11r^2 - 132r$$
$$= 891r^4 + 198r^3 - 1,177r^2 - 132r$$

23. The correct answer is $42x^5 - 21x^4 + 24x^3 + 2x^2 + 8$.

$$\left(7x^2 + 4\right)\left(6x^3 - 3x^2 + 2\right) = \left(7x^2 \times 6x^3\right) + \left[7x^2 \times \left(-3x^2\right)\right] + \left(7x^2 \times 2\right) + \left(4 \times 6x^3\right) + \left[4 \times \left(-3x^2\right)\right] + \left(4 \times 2\right)$$
$$= \left(42x^{2+3}\right) + \left(-21x^{2+2}\right) + \left(14x^2\right) + \left(24x^3\right) + \left(-12x^2\right) + \left(8\right)$$
$$= 42x^5 - 21x^4 + 14x^2 + 24x^3 - 12x^2 + 8$$
$$= 42x^5 - 21x^4 + 24x^3 + 14x^2 - 12x^2 + 8$$
$$= 42x^5 - 21x^4 + 24x^3 + 2x^2 + 8$$

24. The correct answer is $7t^4 - 19t^3 - 35t^2 - 4t$.

$$\left(t^2 - 4t\right)\left(7t^2 + 9t + 1\right) = \left(t^2 \times 7t^2\right) + \left(t^2 \times 9t\right) + \left(t^2 \times 1\right) - \left(4t \times 7t^2\right) - \left(4t \times 9t\right) - \left(4t \times 1\right)$$
$$= \left(7t^{2+2}\right) + \left(9t^{2+1}\right) + \left(t^2\right) - \left(28t^{1+2}\right) - \left(36t^{1+1}\right) - \left(4t \times 1\right)$$
$$= 7t^4 + 9t^3 + t^2 - 28t^3 - 36t^2 - 4t$$
$$= 7t^4 + 9t^3 - 28t^3 + t^2 - 36t^2 - 4t$$
$$= 7t^4 - 19t^3 - 35t^2 - 4t$$

25. The correct answer is $20a^4 - 104a^3 + 98a^2 + 52a - 54$.

$$\left(4a^2 - 2\right)\left(5a^2 - 26a + 27\right) = \left(4a^2 \times 5a^2\right) + \left[4a^2 \times (-26a)\right] + \left(4a^2 \times 27\right) - \left(2 \times 5a^2\right) - \left[2 \times (-26a)\right] - (2 \times 27)$$

$$= \left(20a^{2+2}\right) + \left(-104a^{2+1}\right) + \left(108a^2\right) - \left(10a^2\right) - (-52a) - (54)$$

$$= 20a^4 - 104a^3 + 108a^2 - 10a^2 + 52a - 54$$

$$= 20a^4 - 104a^3 + 98a^2 + 52a - 54$$

Chapter 7

Polynomial Equations

In this chapter, we'll review
the following concepts:

What is a polynomial equation?
Factoring polynomial equations
Solving polynomial equations

What is a polynomial equation?

A polynomial equation is an equation that contains a polynomial expression set equal to zero. Here are some examples:

$$5x + 3 = 0$$
$$2n^2 + 4 = 0$$
$$7y^3 + 2y^2 + 6 = 0$$

Polynomial equations are classified based on the degree of the polynomial expression. Polynomial equations in the first degree are called **linear equations**.

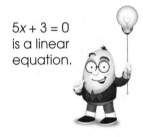

$5x + 3 = 0$
is a linear
equation.

Polynomial equations in the second degree are called **quadratic equations**. The equation $2n^2 + 4 = 0$ is a quadratic equation.

Third-degree polynomial equations are called **cubic equations**. The equation $7y^3 + 2y^2 + 6 = 0$ is a cubic equation, because its largest exponent is 3.

Factoring polynomial equations

We can solve polynomial equations in different ways. One common way of solving these equations is through **factoring**.

To use the factoring method, we must first separate the polynomial expression into its factors. Here we'll focus on factoring binomials and trinomials.

Factoring binomials

Let's start with the equation $3a^2 + 3 = 0$. In this equation, the expression $3a^2 + 3$ is a binomial. It is the sum of two monomial terms.

To factor this binomial, we perform the opposite of distribution. When we distribute, we multiply a term by other terms in parentheses to produce a product:

$$4(x + 6) = 4x^2 + 24$$

In this example, the 4 is distributed over the x and the 6 to produce $4x^2 + 24$.

When we factor, we identify the largest factor that is included in all terms of the polynomial. This is called the **Greatest Common Factor**, or GCF.

In the expression $3a^2 + 3$, the greatest common factor is 3. The number 3 is the only factor that is common to both terms $3a^2$ and 3.

To factor this expression, we separate the binomial into the product of two factors:

$$3a^2 + 3 = 3(a^2 + 1)$$

The factors of the binomial are 3 and $(a^2 + 1)$.

Here is another example where both terms of the binomial contain variables.

$$7n^2 + 7n = 0$$

To factor this equation, we must find the greatest common factor of $7n^2 + 7n$. The number 7 is a factor of both terms, so we can factor 7 out as shown:

$$7n^2 + 7n = 0$$
$$7(n^2 + n) = 0$$

We have factored out the 7, which leaves us with $7(n^2 + n)$. The variable n is common to both n^2 and n, so the expression in parentheses can be factored even further:

$$7n^2 + 7n = 0$$
$$7(n^2 + n) = 0$$
$$7n(n + 1) = 0$$

The greatest common factor of the polynomial expression is $7n$, because it can be factored out of both terms. The factors of the binomial are $7n$ and $(n + 1)$.

Practice Questions—Factoring binomial equations

Directions: Factor the binomials in the equations shown. You will find the Practice Question Solutions on page 153.

1. $4b^2 + b = 0$

2. $p^2 - 12p = 0$

3. $5x + 20 = 0$

4. $s^2 + 16s = 0$

5. $3a^2 + 21a = 0$

6. $4y^2 + 24 = 0$

7. $9m^2 - 18m = 0$

8. $14z - 7 = 0$

9. $s^2 + s = 0$

10. $11b^2 - 66 = 0$

Factoring trinomials

A trinomial equation is an equation containing a trinomial expression, such as $3x^2 + 7x + 4 = 0$. To factor trinomial equations, we can use the reverse of the FOIL method learned in the last chapter.

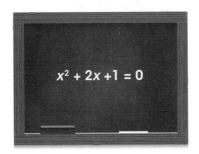

To factor this equation, let's first set out the parentheses we will use for the two factors:

$$x^2 + 2x + 1 = 0$$
$$(\quad)(\quad) = 0$$

We know that the first variable inside each pair of parentheses will be x, because x times x produces x^2. Enter the variables inside the parentheses:

$$x^2 + 2x + 1 = 0$$
$$(x + \underline{\quad})(x + \underline{\quad}) = 0$$

Now, we are looking for two numbers to fill in the blanks in the parentheses. When multiplied together, these numbers must produce a product of 1, the last term in the trinomial:

$$x^2 + 2x + 1 = 0$$
$$(x + 1)(x + 1) = 0$$

The numbers 1 and 1 are the only options. Check the factoring using FOIL:

$$(x + 1)(x + 1) = 0$$
$$x^2 + x + x + 1 = 0$$
$$x^2 + 2x + 1 = 0$$

The factors produce the original expression, so $(x + 1)$ and $(x + 1)$ are correct.

Here's another example using reverse FOIL.

$$5a^2 + a - 6 = 0$$

In this case, the coefficient of the first term is 5. What terms, when multiplied, produce a product of $5a^2$? The only options are $5a$ and a. So, these must be the first terms of each factor:

$$5a^2 + a - 6 = 0$$
$$(5a + \underline{})(a - \underline{}) = 0$$

Now we're looking for the last terms of each factor. When multiplied, they must produce a product of –6. Since the product is a negative number, one of the terms will be negative, and one will be positive. What terms, when multiplied, produce a product of 6? Here we have two options: 6 and 1 or 2 and 3.

The terms 6 and 1 will work. These are the only terms that produce a middle term of a:

$$5a^2 + a - 6 = 0$$
$$(5a + 6)(a - 1) = 0$$

Check the factors by using FOIL to multiply them:

$$(5a + 6)(a - 1) = 0$$
$$5a^2 - 5a + 6a - 6 = 0$$
$$5a^2 + a - 6 = 0$$

The factors multiply to produce the original trinomial, so $(5a + 6)$ and $(a - 1)$ are the factors of the equation.

Practice Questions—Factoring trinomial equations

Directions: Factor the trinomials in the equations shown. You will find the Practice Question Solutions on page 153.

11. $x^2 - x - 6 = 0$

12. $6k^2 - 12k - 48 = 0$

13. $3n^2 - 19n - 14 = 0$

14. $-r^2 + 14r - 49 = 0$

15. $2c^2 + 30c + 52 = 0$

16. $7p^2 + 37p - 30 = 0$

17. $j^2 - 14j + 33 = 0$

18. $-d^2 - 2d + 63 = 0$

19. $c^2 - 24c + 144 = 0$

20. $5n^2 + 8n + 3 = 0$

Solving polynomial equations

Once we have factored a polynomial equation, we can solve it by setting both factors equal to zero. The factors multiply to produce a product of zero, so we know that at least one of the factors is zero.

Equations with binomials

To solve a binomial equation, first factor the equation:

$$13r^2 + 13r = 0$$

This equation can be factored into $13r$ and $(r+1)$:

$$13r^2 + 13r = 0$$
$$13r(r+1) = 0$$

In order for the factors to produce a product of zero, one or both of the factors must equal zero. Set the first factor equal to zero, and solve for the value of r:

$$13r = 0$$
$$\frac{13r}{13} = \frac{0}{13}$$
$$r = \frac{0}{13}$$
$$r = 0$$

The values of r are called the **solutions** of the equation. One value of r is 0. Set the second factor equal to zero, and solve:

$$(r + 1) = 0$$
$$r + 1 - 1 = 0 - 1$$
$$r = 0 - 1$$
$$r = -1$$

Another value of r is –1. The solutions of the equation are 0 and –1.

Practice Questions—Solving binomial equations

Directions: Find the solutions to the equations shown. You will find the Practice Question Solutions on page 154.

21. $6n^2 + 42n = 0$

22. $t^2 - 9t = 0$

23. $-m^2 - 14m = 0$

24. $14r^2 + 7r = 0$

25. $-8a^2 + 56a = 0$

26. $18n^2 - 36n = 0$

27. $-4x^2 + 40x = 0$

28. $-c^2 - 17c = 0$

29. $4p^2 + 16p = 0$

30. $-28t^2 + 84t = 0$

Equations with trinomials

To solve equations with trinomial expressions, we follow a similar process as we did with binomials. First, factor the equation using reverse FOIL. Then, set each factor equal to zero and solve.

$$9m^2 + 9m - 18 = 0$$

This equation can be factored into $(3m + 6)$ and $(3m - 3)$:

$$9m^2 + 9m - 18 = 0$$
$$(3m + 6)(3m - 3) = 0$$

Set each factor equal to zero and solve:

$$3m + 6 = 0 \qquad\qquad 3m - 3 = 0$$
$$3m + 6 - 6 = 0 - 6 \qquad\qquad 3m - 3 + 3 = 0 + 3$$
$$3m = 0 - 6 \qquad\qquad 3m = 0 + 3$$
$$3m = -6 \qquad\qquad 3m = 3$$
$$\frac{3m}{3} = \frac{-6}{3} \qquad\qquad \frac{3m}{3} = \frac{3}{3}$$
$$m = \frac{-6}{3} \qquad\qquad m = \frac{3}{3}$$
$$m = -2 \qquad\qquad m = 1$$

Solving both equations, we see that $m = -2$ and $m = 1$. The solutions of the equation are –2 and 1.

Practice Questions—Solving trinomial equations

Directions: Find the solutions to the equations shown. You will find the Practice Question Solutions on page 158.

31. $w^2 - 9w + 14 = 0$

32. $2q^2 + 10q + 8 = 0$

33. $2m^2 + 18m + 28 = 0$

34. $a^2 + 4a - 21 = 0$

35. $4r^2 - 24r + 20 = 0$

36. $f^2 + 17f + 66 = 0$

37. $-3t^2 + 12t - 9 = 0$

38. $-2b^2 + 22b - 56 = 0$

39. $g^2 + 14g + 49 = 0$

40. $-3m^2 + 3m + 36 = 0$

Chapter Review

Directions: Perform the operations shown. Solutions can be found on page 163.

1. Factor the binomial in the equation shown.

   $s^2 - 3s = 0$

2. Factor the binomial in the equation shown.

   $6m^2 + m = 0$

3. Factor the binomial in the equation shown.

   $7x - 21 = 0$

4. Factor the binomial in the equation shown.

   $3q^2 + 18q = 0$

5. Factor the binomial in the equation shown.

   $-s^2 + 8s = 0$

6. Factor the trinomial in the equation shown.

   $x^2 - 4x - 21 = 0$

7. Factor the trinomial in the equation shown.

   $y^2 - 6y + 5 = 0$

8. Factor the trinomial in the equation shown.

   $3b^2 - 7b - 6 = 0$

9.

   Factor the trinomial equation shown.

   $-h^2 + 10h - 25 = 0$

10. Factor the trinomial in the equation shown.

    $3k^2 - 2k - 1 = 0$

11. Find the solutions to the equation shown.

    $2r^2 - r = 0$

12. Find the solutions to the equation shown.

$3n^2 + 36n = 0$

13. Find the solutions to the equation shown.

$-v^2 - 13v = 0$

14. Find the solutions to the equation shown.

$-6c^2 + 48c = 0$

15. Find the solutions to the equation shown.

$9t^2 - 81t = 0$

16. Find the solutions to the equation shown.

$y^2 + 6y + 5 = 0$

17. Find the solutions to the equation shown.

$n^2 - 13n + 36 = 0$

18. Find the solutions to the equation shown.

$3p^2 + 21p + 18 = 0$

19. Find the solutions to the equation shown.

$a^2 + 2a - 35 = 0$

20. Find the solutions to the equation shown.

$6b^2 - 48b + 42 = 0$

21. Find the solutions to the equation shown.

$h^2 + 16h + 39 = 0$

22. Find the solutions to the equation shown.

$-2t^2 + 8t - 8 = 0$

23. Find the solutions to the equation shown.

$-3r^2 + 27r - 60 = 0$

24. Find the solutions to the equation shown.

$a^2 - 10a + 25 = 0$

25. Find the solutions to the equation shown.

$-2n^2 + 14n + 36 = 0$

**Practice
Question
Solutions**

Factoring binomial equations

1. The factors are b and $(4b + 1)$.

   $$4b^2 + b = 0$$
   $$b(4b + 1) = 0$$

2. The factors are p and $(p - 12)$.

   $$p^2 - 12p = 0$$
   $$p(p - 12) = 0$$

3. The factors are 5 and $(x + 4)$.

   $$5x + 20 = 0$$
   $$5(x + 4) = 0$$

4. The factors are s and $(s + 16)$.

   $$s^2 + 16s = 0$$
   $$s(s + 16) = 0$$

5. The factors are $3a$ and $(a + 7)$.

   $$3a^2 + 21a = 0$$
   $$3(a^2 + 7a) = 0$$
   $$3a(a + 7) = 0$$

6. The factors are 4 and $(y^2 + 6)$.

   $$4y^2 + 24 = 0$$
   $$4(y^2 + 6) = 0$$

7. The factors are $9m$ and $(m - 2)$.

   $$9m^2 - 18m = 0$$
   $$9(m^2 - 2m) = 0$$
   $$9m(m - 2) = 0$$

8. The factors are 7 and $(2z - 1)$.

   $$14z - 7 = 0$$
   $$7(2z - 1) = 0$$

9. The factors are s and $(s + 1)$.

   $$s^2 + s = 0$$
   $$s(s + 1) = 0$$

10. The factors are 11 and $(b^2 - 6)$.

    $$11b^2 - 66 = 0$$
    $$11(b^2 - 6) = 0$$

Factoring trinomial equations

11. The factors are $(x - 3)$ and $(x + 2)$.

    $$x^2 - x - 6 = 0$$
    $$(x - 3)(x + 2) = 0$$

12. The factors are $(6k + 12)$ and $(k - 4)$.

    $$6k^2 - 12k - 48 = 0$$
    $$(6k + 12)(k - 4) = 0$$

13. The factors are $(3n + 2)$ and $(n - 7)$.

    $$3n^2 - 19n - 14 = 0$$
    $$(3n + 2)(n - 7) = 0$$

14. The factors are $(r - 7)$ and $(-r + 7)$.

    $$-r^2 + 14r - 49 = 0$$
    $$(r - 7)(-r + 7) = 0$$

15. The factors are $(2c + 4)$ and $(c + 13)$.

$$2c^2 + 30c + 52 = 0$$
$$(2c + 4)(c + 13) = 0$$

16. The factors are $(7p - 5)$ and $(p + 6)$.

$$7p^2 + 37p - 30 = 0$$
$$(7p - 5)(p + 6) = 0$$

17. The factors are $(j - 3)$ and $(j - 11)$.

$$j^2 - 14j + 33 = 0$$
$$(j - 3)(j - 11) = 0$$

18. The factors are $(-d - 9)$ and $(d - 7)$.

$$-d^2 - 2d + 63 = 0$$
$$(-d - 9)(d - 7) = 0$$

19. The factors are $(c - 12)$ and $(c - 12)$.

$$c^2 - 24c + 144 = 0$$
$$(c - 12)(c - 12) = 0$$

20. The factors are $(5n + 3)$ and $(n + 1)$.

$$5n^2 + 8n + 3 = 0$$
$$(5n + 3)(n + 1) = 0$$

Solving binomial equations

21. The solutions are 0 and –7.

Factor the equation:

$$6n^2 + 42n = 0$$
$$6(n^2 + 7n) = 0$$
$$6n(n + 7) = 0$$

The factors are $6n$ and $(n + 7)$. Set each factor equal to zero and solve:

$$6n(n + 7) = 0$$
$$6n = 0$$
$$\frac{6n}{6} = \frac{0}{6}$$
$$n = 0$$

$$6n(n + 7) = 0$$
$$n + 7 = 0$$
$$n + 7 - 7 = 0 - 7$$
$$n = 0 - 7$$
$$n = -7$$

22. The solutions are 0 and 9.

Factor the equation:

$$t^2 - 9t = 0$$
$$t(t - 9) = 0$$

Set each factor equal to zero and solve:

$$t(t-9) = 0$$
$$t = 0$$

$$t(t-9) = 0$$
$$t - 9 = 0$$
$$t - 9 + 9 = 0 + 9$$
$$t = 0 + 9$$
$$t = 9$$

23. The solutions are 0 and -14.

Factor the equation:

$$-m^2 - 14m = 0$$
$$-m(m+14) = 0$$

The factors are $-m$ and $(m+14)$. Set each factor equal to zero and solve:

$$-m(m+14) = 0$$
$$-m = 0$$
$$\frac{-m}{-1} = \frac{0}{-1}$$
$$m = 0$$

$$-m(m+14) = 0$$
$$m + 14 = 0$$
$$m + 14 - 14 = 0 - 14$$
$$m = 0 - 14$$
$$m = -14$$

24. The solutions are 0 and $-\frac{1}{2}$.

Factor the equation:

$$14r^2 + 7r = 0$$
$$7r(2r+1) = 0$$

The factors are $7r$ and $(2r+1)$. Set each factor equal to zero and solve:

$$7r(2r+1) = 0$$
$$7r = 0$$
$$\frac{7r}{7} = \frac{0}{7}$$
$$r = 0$$

$$7r(2r+1) = 0$$
$$2r + 1 = 0$$
$$2r + 1 - 1 = 0 - 1$$
$$2r = 0 - 1$$
$$2r = -1$$
$$\frac{2r}{2} = \frac{-1}{2}$$
$$r = -\frac{1}{2}$$

25. The solutions are 0 and 7.

Factor the equation:

$$-8a^2 + 56a = 0$$
$$-8(a^2 - 7a) = 0$$
$$-8a(a - 7) = 0$$

The factors are $-8a$ and $(a - 7)$. Set each factor equal to zero and solve:

$$-8a(a - 7) = 0 \qquad\qquad -8a(a - 7) = 0$$
$$-8a = 0 \qquad\qquad\qquad a - 7 = 0$$
$$\frac{-8a}{-8} = \frac{0}{-8} \qquad\qquad a - 7 + 7 = 0 + 7$$
$$a = \frac{0}{-8} \qquad\qquad\qquad a = 0 + 7$$
$$a = 0 \qquad\qquad\qquad\quad a = 7$$

26. The solutions are 0 and 2.

Factor the equation:

$$18n^2 - 36n = 0$$
$$18(n^2 - 2n) = 0$$
$$18n(n - 2) = 0$$

The factors are $18n$ and $(n - 2)$. Set each factor equal to zero and solve:

$$18n(n - 2) = 0 \qquad\qquad 8n(n - 2) = 0$$
$$18n = 0 \qquad\qquad\qquad n - 2 = 0$$
$$\frac{18n}{18} = \frac{0}{18} \qquad\qquad n - 2 + 2 = 0 + 2$$
$$n = 0 \qquad\qquad\qquad\quad n = 0 + 2$$
$$\qquad\qquad\qquad\qquad\qquad n = 2$$

27. The solutions are 0 and 10.

Factor the equation:

$$-4x^2 + 40x = 0$$
$$-4(x^2 - 10x) = 0$$
$$-4x(x - 10) = 0$$

The factors are $-4x$ and $(x-10)$. Set each factor equal to zero and solve:

$$-4x(x-10) = 0$$
$$-4x = 0$$
$$\frac{-4x}{-4} = \frac{0}{-4}$$
$$x = 0$$

$$-4x(x-10) = 0$$
$$x - 10 = 0$$
$$x + 10 + 10 = 0 + 10$$
$$x = 0 + 10$$
$$x = 10$$

28. The solutions are 0 and -17.

Factor the equation:

$$-c^2 - 17c = 0$$
$$-c(c+17) = 0$$

The factors are $-c$ and $(c+17)$. Set each factor equal to zero and solve:

$$-c(c+17) = 0$$
$$-c = 0$$
$$\frac{-c}{-1} = \frac{0}{-1}$$
$$c = 0$$

$$-c(c+17) = 0$$
$$c + 17 = 0$$
$$c + 17 - 17 = 0 - 17$$
$$c = 0 - 17$$
$$c = -17$$

29. The solutions are 0 and -4.

Factor the equation:

$$4p^2 + 16p = 0$$
$$4(p^2 + 4p) = 0$$
$$4p(p + 4) = 0$$

The factors are $4p$ and $(p+4)$. Set each factor equal to zero and solve:

$$4p(p+4) = 0$$
$$4p = 0$$
$$\frac{4p}{4} = \frac{0}{4}$$
$$p = 0$$

$$4p(p+4) = 0$$
$$p + 4 = 0$$
$$p + 4 - 4 = 0 - 4$$
$$p = 0 - 4$$
$$p = -4$$

30. The solutions are 0 and 3.

Factor the equation:

$$-28t^2 + 84t = 0$$
$$-28(t^2 - 3t) = 0$$
$$-28t(t - 3) = 0$$

The factors are $-28t$ and $(t - 3)$. Set each factor equal to zero and solve:

$$-28t(t - 3) = 0$$
$$-28t = 0$$
$$\frac{-28t}{-28} = \frac{0}{-28}$$
$$t = 0$$

$$-28t(t - 3) = 0$$
$$t - 3 = 0$$
$$t - 3 + 3 = 0 + 3$$
$$t = 0 + 3$$
$$t = 3$$

Solving trinomial equations

31. The solutions are 7 and 2.

Factor the equation:

$$w^2 - 9w + 14 = 0$$
$$(w - 7)(w - 2) = 0$$

The factors are $(w - 7)$ and $(w - 2)$. Set each factor equal to zero and solve:

$$w - 7 = 0$$
$$w - 7 + 7 = 0 + 7$$
$$w = 0 + 7$$
$$w = 7$$

$$w - 2 = 0$$
$$w - 2 + 2 = 0 + 2$$
$$w = 0 + 2$$
$$w = 2$$

32. The solutions are -1 and -4.

Factor the equation:

$$2q^2 + 10q + 8 = 0$$
$$(2q + 2)(q + 4) = 0$$

The factors are $(2q + 2)$ and $(q + 4)$. Set each factor equal to zero and solve:

$$2q + 2 = 0$$
$$2q + 2 - 2 = 0 - 2$$
$$2q = 0 - 2$$
$$2q = -2$$
$$\frac{2q}{2} = \frac{-2}{2}$$
$$q = \frac{-2}{2}$$
$$q = -1$$

$$q + 4 = 0$$
$$q + 4 - 4 = 0 - 4$$
$$q = 0 - 4$$
$$q = -4$$

33.

The solutions are –2 and –7.

Factor the equation:

$2m^2 + 18m + 28 = 0$

$(2m + 4)(m + 7) = 0$

The factors are $(2m + 4)$ and $(m + 7)$. Set each factor equal to zero and solve:

$2m + 4 = 0$ $m + 7 = 0$

$2m + 4 - 4 = 0 - 4$ $m + 7 - 7 = 0 - 7$

$2m = 0 - 4$ $m = 0 - 7$

$2m = -4$ $m = -7$

$\dfrac{2m}{2} = \dfrac{-4}{2}$

$m = \dfrac{-4}{2}$

$m = -2$

34. The solutions are 3 and –7.

Factor the equation:

$a^2 + 4a - 21 = 0$

$(a - 3)(a + 7) = 0$

The factors are and $(a - 3)$ and $(a + 7)$. Set each factor equal to zero and solve:

$a - 3 = 0$ $a + 7 = 0$

$a - 3 + 3 = 0 + 3$ $a + 7 - 7 = 0 - 7$

$a = 0 + 3$ $a = 0 - 7$

$a = 3$ $a = -7$

35. The solutions are 1 and 5.

Factor the equation:

$$4r^2 - 24r + 20 = 0$$
$$(4r - 4)(r - 5) = 0$$

The factors are $(4r - 4)$ and $(r - 5)$. Set each factor equal to zero and solve:

$$4r - 4 = 0 \qquad\qquad r - 5 = 0$$
$$4r - 4 + 4 = 0 + 4 \qquad\qquad r - 5 + 5 = 0 + 5$$
$$4r = 0 + 4 \qquad\qquad r = 0 + 5$$
$$4r = 4 \qquad\qquad r = 5$$
$$\frac{4r}{4} = \frac{4}{4}$$
$$r = \frac{4}{4}$$
$$r = 1$$

36. The solutions are –11 and –6.

Factor the equation:

$$f^2 + 17f + 66 = 0$$
$$(f + 11)(f + 6) = 0$$

The factors are $(f + 11)$ and $(f + 6)$. Set each factor equal to zero and solve:

$$f + 11 = 0 \qquad\qquad f + 6 = 0$$
$$f + 11 - 11 = 0 - 11 \qquad\qquad f + 6 - 6 = 0 - 6$$
$$f = 0 - 11 \qquad\qquad f = 0 - 6$$
$$f = -11 \qquad\qquad f = -6$$

37. The solutions are 3 and 1.

Factor the equation:

$$-3t^2 + 12t - 9 = 0$$
$$(-3t + 9)(t - 1) = 0$$

The factors are $(-3t + 9)$ and $(t - 1)$. Set each factor equal to zero and solve:

$$-3t + 9 = 0$$
$$-3t + 9 - 9 = 0 - 9$$
$$-3t = 0 - 9$$
$$-3t = -9$$
$$\frac{-3t}{-3} = \frac{-9}{-3}$$
$$t = \frac{-9}{-3}$$
$$t = 3$$

$$t - 1 = 0$$
$$t - 1 + 1 = 0 + 1$$
$$t = 0 + 1$$
$$t = 1$$

38. The solutions are 4 and 7.

Factor the equation:

$$-2b^2 + 22b - 56 = 0$$
$$(2b - 8)(-b + 7) = 0$$

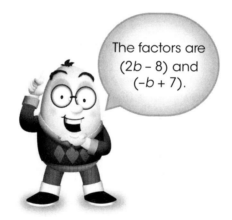

The factors are $(2b - 8)$ and $(-b + 7)$.

Set each factor equal to zero and solve:

$$2b - 8 = 0$$
$$2b - 8 + 8 = 0 + 8$$
$$2b = 0 + 8$$
$$2b = 8$$
$$\frac{2b}{2} = \frac{8}{2}$$
$$b = \frac{8}{2}$$
$$b = 4$$

$$-b + 7 = 0$$
$$-b + 7 - 7 = 0 - 7$$
$$-b = 0 - 7$$
$$-b = -7$$
$$\frac{-b}{-1} = \frac{-7}{-1}$$
$$b = \frac{-7}{-1}$$
$$b = 7$$

39. The solution is –7.

Factor the equation:

$$g^2 + 14g + 49 = 0$$
$$(g + 7)(g + 7) = 0$$

The factors are $(g + 7)$ and $(g + 7)$. The factors are identical, so we set one factor equal to zero and solve:

$$g + 7 = 0$$
$$g + 7 - 7 = 0 - 7$$
$$g = 0 - 7$$
$$g = -7$$

40. The solutions are –3 and 4.

Factor the equation:

$$-3m^2 + 3m + 36 = 0$$
$$(3m + 9)(-m + 4) = 0$$

The factors are $(3m + 9)$ and $(-m + 4)$. Set each factor equal to zero and solve:

$$3m + 9 = 0 \qquad\qquad -m + 4 = 0$$
$$3m + 9 - 9 = 0 - 9 \qquad -m + 4 - 4 = 0 - 4$$
$$3m = 0 - 9 \qquad\qquad -m = 0 - 4$$
$$3m = -9 \qquad\qquad -m = -4$$
$$\frac{3m}{3} = \frac{-9}{3} \qquad\qquad \frac{-m}{-1} = \frac{-4}{-1}$$
$$m = \frac{-9}{3} \qquad\qquad m = \frac{-4}{-1}$$
$$m = -3 \qquad\qquad m = 4$$

**Chapter
Review
Solutions**

1. The factors of the binomial are s and $(s-3)$.

$$s^2 - 3s = 0$$
$$s(s-3) = 0$$

2. The factors of the binomial are m and $(6m+1)$.

$$6m^2 + m = 0$$
$$m(6m+1) = 0$$

3. The factors of the binomial are 7 and $(x-3)$.

$$7x - 21 = 0$$
$$7(x-3) = 0$$

4. The factors of the binomial are $3q$ and $(q+6)$.

$$3q^2 + 18q = 0$$
$$3(q^2 + 6q) = 0$$
$$3q(q+6) = 0$$

5. The factors of the binomial are s and $(-s+8)$.

$$-s^2 + 8s = 0$$
$$s(-s+8) = 0$$

6. The factors of the trinomial are $(x+3)$ and $(x-7)$.

$$x^2 - 4x - 21 = 0$$
$$(x+3)(x-7) = 0$$

7. The factors of the trinomial are $(y-5)$ and $(y-1)$.

$$y^2 - 6y + 5 = 0$$
$$(y-5)(y-1) = 0$$

8. The factors of the trinomial are $(3b+2)$ and $(b-3)$.

$$3b^2 - 7b - 6 = 0$$
$$(3b+2)(b-3) = 0$$

9. The factors of the trinomial are $(-h+5)$ and $(h-5)$.

$$-h^2 + 10h - 25 = 0$$
$$(-h+5)(h-5) = 0$$

10. The factors of the trinomial are $(k-1)$ and $(3k+1)$.

$$3k^2 - 2k - 1 = 0$$
$$(k-1)(3k+1) = 0$$

11.

The solutions are 0 and $\frac{1}{2}$.

Factor the equation:

$2r^2 - r = 0$
$r(2r - 1) = 0$

The factors are r and $(2r - 1)$. Set each factor equal to zero and solve:

$r(2r - 1) = 0$ $r(2r - 1) = 0$
$\qquad\quad r = 0$ $\qquad 2r - 1 = 0$
$\qquad\qquad\qquad\qquad\qquad 2r - 1 + 1 = 0 + 1$
$\qquad\qquad\qquad\qquad\qquad\quad 2r = 1$
$\qquad\qquad\qquad\qquad\qquad\quad \frac{2r}{2} = \frac{1}{2}$
$\qquad\qquad\qquad\qquad\qquad\qquad r = \frac{1}{2}$

12. The solutions are 0 and –12.

Factor the equation:

$3n^2 + 36n = 0$
$3(n^2 + 12n) = 0$
$3n(n + 12) = 0$

The factors are $3n$ and $(n + 12)$. Set each factor equal to zero and solve:

$$3n(n + 12) = 0$$
$$3n = 0$$
$$\frac{3n}{3} = \frac{0}{3}$$
$$n = \frac{0}{3}$$
$$n = 0$$

$$3n(n + 12) = 0$$
$$n + 12 = 0$$
$$n + 12 - 12 = 0 - 12$$
$$n = 0 - 12$$
$$n = -12$$

13. The solutions are 0 and –13.

Factor the equation:

$$-v^2 - 13v = 0$$
$$-v(v + 13) = 0$$

The factors are $-v$ and $(v + 13)$. Set each factor equal to zero and solve:

$$-v(v + 13) = 0$$
$$-v = 0$$
$$\frac{-v}{-1} = \frac{0}{-1}$$
$$v = \frac{0}{-1}$$
$$v = 0$$

$$-v(v + 13) = 0$$
$$v + 13 = 0$$
$$v + 13 - 13 = 0 - 13$$
$$v = 0 - 13$$
$$v = -13$$

14. Factor the equation:

$$-6c^2 + 48c = 0$$
$$-6(c^2 - 8c) = 0$$
$$-6c(c - 8) = 0$$

The factors are $-6c$ and $(c - 8)$. Set each factor equal to zero and solve:

$$-6c(c - 8) = 0$$
$$-6c = 0$$
$$\frac{-6c}{-6} = \frac{0}{-6}$$
$$c = \frac{0}{-6}$$
$$c = 0$$

$$-6c(c - 8) = 0$$
$$c - 8 = 0$$
$$c - 8 + 8 = 0 + 8$$
$$c = 0 + 8$$
$$c = 8$$

The solutions are 0 and 8.

15. The solutions are 0 and 9.

Factor the equation:

$$9t^2 - 81t = 0$$
$$9(t^2 - 9t) = 0$$
$$9t(t - 9) = 0$$

The factors are $9t$ and $(t - 9)$. Set each factor equal to zero and solve:

$$9t(t - 9) = 0 \qquad\qquad 9t(t - 9) = 0$$
$$9t = 0 \qquad\qquad\qquad t - 9 = 0$$
$$\frac{9t}{9} = \frac{0}{9} \qquad\qquad t - 9 + 9 = 0 + 9$$
$$t = \frac{0}{9} \qquad\qquad\qquad t = 0 + 9$$
$$t = 0 \qquad\qquad\qquad\quad t = 9$$

16. The solutions are –1 and –5.

Factor the equation:

$$y^2 + 6y + 5 = 0$$
$$(y + 1)(y + 5) = 0$$

The factors are $(y + 1)$ and $(y + 5)$. Set each factor equal to zero and solve:

$$y + 1 = 0 \qquad\qquad\qquad y + 5 = 0$$
$$y + 1 - 1 = 0 - 1 \qquad\qquad y + 5 - 5 = 0 - 5$$
$$y = 0 - 1 \qquad\qquad\qquad y = 0 - 5$$
$$y = -1 \qquad\qquad\qquad\quad y = -5$$

17. The solutions are 9 and 4.

Factor the equation:

$$n^2 - 13n + 36 = 0$$
$$(n - 9)(n - 4) = 0$$

The factors are $(n - 9)$ and $(n - 4)$. Set each factor equal to zero and solve:

$$n - 9 = 0 \qquad\qquad\qquad n - 4 = 0$$
$$n - 9 + 9 = 0 + 9 \qquad\qquad n - 4 + 4 = 0 + 4$$
$$n = 0 + 9 \qquad\qquad\qquad n = 0 + 4$$
$$n = 9 \qquad\qquad\qquad\qquad n = 4$$

18. The solutions are –1 and –6.

Factor the equation:

$$3p^2 + 21p + 18 = 0$$
$$(3p + 3)(p + 6) = 0$$

The factors are $(3p + 3)$ and $(p + 6)$. Set each factor equal to zero and solve:

$$3p + 3 = 0 \qquad\qquad p + 6 = 0$$
$$3p + 3 - 3 = 0 - 3 \qquad\qquad p + 6 - 6 = 0 - 6$$
$$3p = 0 - 3 \qquad\qquad p = 0 - 6$$
$$3p = -3 \qquad\qquad p = -6$$
$$\frac{3p}{3} = \frac{-3}{3}$$
$$p = \frac{-3}{3}$$
$$p = -1$$

19.

The solutions are 5 and –7.

Factor the equation:

$$a^2 + 2a - 35 = 0$$
$$(a - 5)(a + 7) = 0$$

The factors are $(a - 5)$ and $(a + 7)$. Set each factor equal to zero and solve:

$$a - 5 = 0 \qquad\qquad a + 7 = 0$$
$$a - 5 + 5 = 0 + 5 \qquad\qquad a + 7 - 7 = 0 - 7$$
$$a = 0 + 5 \qquad\qquad a = 0 - 7$$
$$a = 5 \qquad\qquad a = -7$$

20. The solutions are 1 and 7.

Factor the equation:

$6b^2 - 48b + 42 = 0$

$(6b - 6)(b - 7) = 0$

The factors are $(6b - 6)$ and $(b - 7)$. Set each factor equal to zero and solve:

$$6b - 6 = 0$$
$$6b - 6 + 6 = 0 + 6$$
$$6b = 0 + 6$$
$$6b = 6$$
$$\frac{6b}{6} = \frac{6}{6}$$
$$b = \frac{6}{6}$$
$$b = 1$$

$$b - 7 = 0$$
$$b - 7 + 7 = 0 + 7$$
$$b = 0 + 7$$
$$b = 7$$

21. The solutions are –13 and –3.

Factor the equation:

$h^2 + 16h + 39 = 0$

$(h + 13)(h + 3) = 0$

The factors are $(h - 13)$ and $(h + 3)$. Set each factor equal to zero and solve:

$$h + 13 = 0$$
$$h + 13 - 13 = 0 - 13$$
$$h = 0 - 13$$
$$h = -13$$

$$h + 3 = 0$$
$$h + 3 - 3 = 0 - 3$$
$$h = 0 - 3$$
$$h = -3$$

22. Factor the equation:

$-2t^2 + 8t - 8 = 0$

$(-2t + 4)(t - 2) = 0$

The solution is 2.

The factors are $(-2t + 4)$ and $(t - 2)$. Set each factor equal to zero and solve:

$$-2t + 4 = 0$$
$$-2t + 4 - 4 = 0 - 4$$
$$-2t = 0 - 4$$
$$-2t = -4$$
$$\frac{-2t}{-2} = \frac{-4}{-2}$$
$$t = \frac{-4}{-2}$$
$$t = 2$$

$$t - 2 = 0$$
$$t - 2 + 2 = 0 + 2$$
$$t = 0 + 2$$
$$t = 2$$

23. The solutions are 4 and 5.

Factor the equation:

$$-3r^2 + 27r - 60 = 0$$
$$(3r - 12)(-r + 5) = 0$$

The factors are $(3r - 12)$ and $(-r + 5)$. Set each factor equal to zero and solve:

$$3r - 12 = 0$$
$$3r - 12 + 12 = 0 + 12$$
$$3r = 0 + 12$$
$$3r = 12$$
$$\frac{3r}{3} = \frac{12}{3}$$
$$r = \frac{12}{3}$$
$$r = 4$$

$$-r + 5 = 0$$
$$-r + 5 - 5 = 0 - 5$$
$$-r = 0 - 5$$
$$-r = -5$$
$$\frac{-r}{-1} = \frac{-5}{-1}$$
$$r = \frac{-5}{-1}$$
$$r = 5$$

24. The solution is 5.

Factor the equation:

$$a^2 - 10a + 25 = 0$$
$$(a - 5)(a - 5) = 0$$

The factors are $(a - 5)$ and $(a - 5)$. The factors are identical, so we can set one factor equal to zero and solve:

$$a - 5 = 0$$
$$a - 5 + 5 = 0 + 5$$
$$a = 0 + 5$$
$$a = 5$$

25. The solutions are –2 and 9.

Factor the equation:

$-2n^2 + 14n + 36 = 0$

$(2n + 4)(-n + 9) = 0$

The factors are $(2n + 4)$ and $(-n + 9)$. Set each factor equal to zero and solve:

$$2n + 4 = 0 \qquad\qquad -n + 9 = 0$$
$$2n + 4 - 4 = 0 - 4 \qquad\qquad -n + 9 - 9 = 0 - 9$$
$$2n = 0 - 4 \qquad\qquad -n = 0 - 9$$
$$2n = -4 \qquad\qquad -n = -9$$
$$\frac{2n}{2} = \frac{-4}{2} \qquad\qquad \frac{-n}{-1} = \frac{-9}{-1}$$
$$n = \frac{-4}{2} \qquad\qquad n = \frac{-9}{-1}$$
$$n = -2 \qquad\qquad n = 9$$

Chapter 8

Absolute Value

In this chapter, we'll review the following concepts:

What is absolute value?
Absolute value with variables
Solving absolute value equations

What is absolute value?

The absolute value of a number is the distance that number lies from zero on a number line.

The absolute value of 7 is 7.

The number 7 lies 7 units from zero on a number line.

The absolute value of 3 is 3. The number 3 lies 3 units from zero on a number line.

Absolute value works the same way with fractions as with whole numbers.

The absolute value of $\frac{1}{2}$ is $\frac{1}{2}$. The fraction $\frac{1}{2}$ lies exactly one-half unit from zero on the number line.

We can determine the absolute value of negative numbers as well. The absolute value of –6, for instance, is 6. The number –6 lies 6 units away from zero on the number line, in the negative direction.

The symbol for absolute value is two vertical bars:

$$|5| = 5$$

This equation shows that the absolute value of 5 is 5.

$$|-30| = 30$$

This equation shows that the absolute value of –30 is 30.

Absolute value with variables

We use the concept of absolute value with variables as well as with numbers. When you see a variable in absolute value notation, you can solve for the value of the variable.

Example

In the following equation, we are given the absolute value of x:

$$|x| = 15$$

To determine the value of x, consider both positive and negative options. The value of x could be 15:

$$|15| = 15$$

This is a true equation, so 15 is one value of x.

The variable x could also have a negative value. If x is –15, the absolute value equation still holds true:

$$|-15| = 15$$

Therefore, the value of x is 15 or –15.

Here are some questions to practice this concept.

Practice Questions—Absolute value with variables

Directions: Find the solutions to the absolute value equations shown. You will find the Practice Question Solutions on page 180.

1. $|c| = 4$

HINT: Remember both positive and negative options.

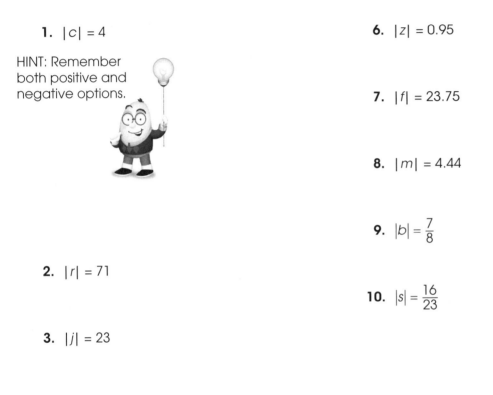

2. $|r| = 71$

3. $|j| = 23$

4. $|x| = 664$

5. $|q| = 5,430$

6. $|z| = 0.95$

7. $|f| = 23.75$

8. $|m| = 4.44$

9. $|b| = \dfrac{7}{8}$

10. $|s| = \dfrac{16}{23}$

Solving absolute value equations

Some algebra equations contain variables plus numbers in absolute value bars. Here is an example:

To solve these types of absolute value equations, we must again consider the positive and negative values of the expression inside the bars.

Examples

In the case of the equation above, there are two possible values for $2 + x$. The absolute value of $2 + x$ is 9. So, $2 + x$ could equal 9 or –9.

To determine the value of x, we must examine both options. First, set $2 + x$ equal to 9 and solve:

$$|2 + x| = 9$$
$$2 + x = 9$$
$$2 - 2 + x = 9 - 2$$
$$x = 9 - 2$$
$$x = 7$$

One value of x is 7. Next, set $2 + x$ equal to –9 and solve:

$$|2 + x| = 9$$
$$2 + x = -9$$
$$2 - 2 + x = -9 - 2$$
$$x = -11$$

Absolute value equations often have two solutions.

Another value of x is –11. The correct answer is $x = 7$ or $x = -11$.

Here's an example involving multiplication.

To find the value of s, create two equations: $4 \times 3s = 24$ and $4 \times 3s = -24$. Then solve for s in both.

$$|4 \times 3s| = 24$$
$$4 \times 3s = 24$$
$$4(3s) = 24$$
$$\frac{4(3s)}{4} = \frac{24}{4}$$
$$3s = \frac{24}{4}$$
$$3s = 6$$

Divide both sides by 3 to find the value of s:

$$3s = 6$$
$$\frac{3s}{3} = \frac{6}{3}$$
$$s = \frac{6}{3}$$
$$s = 2$$

One value of s is 2. Now, solve the second equation:

$$|4 \times 3s| = -24$$
$$4 \times 3s = -24$$
$$4(3s) = -24$$
$$\frac{4(3s)}{4} = \frac{-24}{4}$$
$$3s = \frac{-24}{4}$$
$$3s = -6$$
$$\frac{3s}{3} = \frac{-6}{3}$$
$$s = \frac{-6}{3}$$
$$s = -2$$

The solutions are $s = 2$ or $s = -2$.

Practice Questions—Solving absolute value equations

Directions: Find the solutions to the absolute value equations shown. You will find the Practice Question Solutions on page 180.

11. $|x + 1| = 2$

12. $|c - 2| = 5$

13. $|p + 10| = 7$

14. $|k - 8| = 15$

15. $|h + 4| = 20$

16. $|5y - 3| = 7$

17. $|f \div 2| = 17$

18. $|3r + 9| = 39$

19. $|6q \times 2| = 10$

20. $|2j \div 10| = 8$

Chapter Review

Directions: Find the solutions to the absolute value equations shown. Solutions can be found on page 181.

1. $|c| = 14$

2. $|x| = \frac{11}{12}$

3. $|g| = \frac{2}{3}$

4. $|w| = 404$

5. $|u| = \frac{5}{8}$

6. $|h + 2| = 4$

7. $|b - 8| = 10$

8. $|r + 10| = 20$

9. $|q - 5| = 4$

10. $|k + 11| = 25$

11. $|d - 8| = 45$

12. $|a + 1| = 3$

13. $|j - 9| = 10$

14. $|s + 5| = 7$

15. $|f - 4| = 8$

16. $|7k - 4| = 10$

17. $|2d + 8| = 24$

18. $|t \div 5| = 20$

19. $|2y - 4| = 10$

20. $|3x \div 3| = 18$

21. $|4m \times 8| = 16$

22. $|e \div 4| = 28$

23. $|6u \div 18| = 3$

24. $|9c \times 2| = 10$

25. $|4v + 5| = 45$

Practice Question Solutions

Absolute value with variables

1. The solutions are $c = 4$ or $c = -4$.

2. The solutions are $r = 71$ or $r = -71$.

3. The solutions are $j = 23$ or -23.

4. The solutions are $x = 664$ or $x = -664$.

5. The solutions are $q = 5,430$ or $q = -5,430$.

6. The solutions are $z = 0.95$ or $z = -0.95$.

7. The solutions are $f = 23.75$ or $f = -23.75$.

8. The solutions are $m = 4.44$ or $m = -4.44$.

9. The solutions are $b = \frac{7}{8}$ or $b = -\frac{7}{8}$.

10. The solutions are $s = \frac{16}{23}$ or $s = -\frac{16}{23}$.

Solving absolute value equations

11. The solutions are $x = 1$ or $x = -3$.

$$|x + 1| = 2 \qquad\qquad |x + 1| = 2$$
$$x + 1 = 2 \qquad\qquad x + 1 = -2$$
$$x + 1 - 1 = 2 - 1 \qquad x + 1 - 1 = -2 - 1$$
$$x = 2 - 1 \qquad\qquad x = -2 - 1$$
$$x = 1 \qquad\qquad\quad x = -3$$

12. The solutions are $c = 7$ or $c = -3$.

$$|c - 2| = 5 \qquad\qquad |c - 2| = 5$$
$$c - 2 = 5 \qquad\qquad c - 2 = -5$$
$$c - 2 + 2 = 5 + 2 \qquad c - 2 + 2 = -5 + 2$$
$$c = 5 + 2 \qquad\qquad c = -5 + 2$$
$$c = 7 \qquad\qquad\quad c = -3$$

13. The solutions are $p = -3$ or $p = -17$.

$$|p + 10| = 7 \qquad\qquad |p + 10| = 7$$
$$p + 10 = 7 \qquad\qquad p + 10 = -7$$
$$p + 10 - 10 = 7 - 10 \qquad p + 10 - 10 = -7 - 10$$
$$p = 7 - 10 \qquad\qquad p = -17$$
$$p = -3$$

14. The solutions are $k = 23$ or $k = -7$.

$$|k - 8| = 15 \qquad\qquad |k - 8| = 15$$
$$k - 8 = 15 \qquad\qquad k - 8 = -15$$
$$k - 8 + 8 = 15 + 8 \qquad k - 8 + 8 = -15 + 8$$
$$k = 15 + 8 \qquad\qquad k = -15 + 8$$
$$k = 23 \qquad\qquad\quad k = -7$$

15. The solutions are $h = 16$ or $h = -24$.

$$|h + 4| = 20 \qquad\qquad |h + 4| = 20$$
$$h + 4 = 20 \qquad\qquad h + 4 = -20$$
$$h + 4 - 4 = 20 - 4 \qquad h + 4 - 4 = -20 - 4$$
$$h = 20 - 4 \qquad\qquad h = -20 - 4$$
$$h = 16 \qquad\qquad h = -24$$

16. The solutions are $y = 2$ or $y = -\dfrac{4}{5}$.

$$|5y - 3| = 7 \qquad\qquad |5y - 3| = 7$$
$$5y - 3 = 7 \qquad\qquad 5y - 3 = -7$$
$$5y - 3 + 3 = 7 + 3 \qquad 5y - 3 + 3 = -7 + 3$$
$$5y = 10 \qquad\qquad 5y = -4$$
$$\frac{5y}{5} = \frac{10}{5} \qquad\qquad \frac{5y}{5} = \frac{-4}{5}$$
$$y = 2 \qquad\qquad y = -\frac{4}{5}$$

17. The solutions are $f = 34$ or $f = -34$.

$$|f \div 2| = 17 \qquad\qquad |f \div 2| = 17$$
$$f \div 2 = 17 \qquad\qquad f \div 2 = -17$$
$$2 \times \left(\frac{f}{2}\right) = 17 \times 2 \qquad 2 \times \left(\frac{f}{2}\right) = -17 \times 2$$
$$f = 17 \times 2 \qquad\qquad f = -17 \times 2$$
$$f = 34 \qquad\qquad f = -34$$

18. The solutions are $r = 10$ or $r = -16$.

$$|3r + 9| = 39 \qquad\qquad |3r + 9| = 39$$
$$3r + 9 = 39 \qquad\qquad 3r + 9 = -39$$
$$3r + 9 - 9 = 39 - 9 \qquad 3r + 9 - 9 = -39 - 9$$
$$3r = 30 \qquad\qquad 3r = -48$$
$$\frac{3r}{3} = \frac{30}{3} \qquad\qquad \frac{3r}{3} = \frac{-48}{3}$$
$$r = 10 \qquad\qquad r = -16$$

19. The solutions are $q = \dfrac{5}{6}$ or $q = -\dfrac{5}{6}$.

$$|6q \times 2| = 10 \qquad\qquad |6q \times 2| = 10$$
$$6q \times 2 = 10 \qquad\qquad 6q \times 2 = -10$$
$$2(6q) = 10 \qquad\qquad 2(6q) = -10$$
$$\frac{2(6q)}{2} = \frac{10}{2} \qquad\qquad \frac{2(6q)}{2} = \frac{-10}{2}$$
$$6q = 5 \qquad\qquad 6q = -5$$
$$\frac{6q}{6} = \frac{5}{6} \qquad\qquad \frac{6q}{6} = \frac{-5}{6}$$
$$q = \frac{5}{6} \qquad\qquad q = -\frac{5}{6}$$

20. The solutions are $j = 40$ or $j = -40$.

$$|2j \div 10| = 8 \qquad\qquad |2j \div 10| = 8$$
$$\frac{2j}{10} = 8 \qquad\qquad \frac{2j}{10} = -8$$
$$10 \times \left(\frac{2j}{10}\right) = 8 \times 10 \qquad 10 \times \left(\frac{2j}{10}\right) = -8 \times 10$$
$$2j = 80 \qquad\qquad 2j = -80$$
$$\frac{2j}{2} = \frac{80}{2} \qquad\qquad \frac{2j}{2} = \frac{-80}{2}$$
$$j = 40 \qquad\qquad j = -40$$

Chapter Review Solutions

1. The solutions are $c = 14$ or $c = -14$.

2. The solutions are $x = \dfrac{11}{12}$ or $x = -\dfrac{11}{12}$.

3. The solutions are $g = \dfrac{2}{3}$ or $g = -\dfrac{2}{3}$.

4. The solutions are $w = 404$ or $w = -404$.

5. The solutions are $u = \dfrac{5}{8}$ or $u = -\dfrac{5}{8}$.

6. The solutions are $h = 2$ or $h = -6$.

$$|h + 2| = 4 \qquad\qquad |h + 2| = 4$$
$$h + 2 = 4 \qquad\qquad h + 2 = -4$$
$$h + 2 - 2 = 4 - 2 \qquad h + 2 - 2 = -4 - 2$$
$$h = 4 - 2 \qquad\qquad h = -4 - 2$$
$$h = 2 \qquad\qquad h = -6$$

7. The solutions are $b = 18$ or $b = -2$.

$$|b - 8| = 10 \qquad\qquad |b - 8| = 10$$
$$b - 8 = 10 \qquad\qquad b - 8 = -10$$
$$b - 8 + 8 = 10 + 8 \qquad b - 8 + 8 = -10 + 8$$
$$b = 10 + 8 \qquad\qquad b = -10 + 8$$
$$b = 18 \qquad\qquad b = -2$$

8. The solutions are $r = 10$ or $r = -30$.

$$|r + 10| = 20 \qquad\qquad |r + 10| = 20$$
$$r + 10 = 20 \qquad\qquad r + 10 = -20$$
$$r + 10 - 10 = 20 - 10 \qquad r + 10 - 10 = -20 - 10$$
$$r = 20 - 10 \qquad\qquad r = -20 - 10$$
$$r = 10 \qquad\qquad r = -30$$

9. The solutions are $q = 9$ or $q = 1$.

$$|q - 5| = 4 \qquad\qquad |q - 5| = 4$$
$$q - 5 = 4 \qquad\qquad q - 5 = -4$$
$$q - 5 + 5 = 4 + 5 \qquad q - 5 + 5 = -4 + 5$$
$$q = 4 + 5 \qquad\qquad q = -4 + 5$$
$$q = 9 \qquad\qquad q = 1$$

10. The solutions are $k = 14$ or $k = -36$.

$$|k + 11| = 25 \qquad\qquad |k + 11| = 25$$
$$k + 11 = 25 \qquad\qquad k + 11 = -25$$
$$k + 11 - 11 = 25 - 11 \qquad k + 11 - 11 = -25 - 11$$
$$k = 25 - 11 \qquad\qquad k = -25 - 11$$
$$k = 14 \qquad\qquad k = -36$$

11. The solutions are $d = 53$ or $d = -37$.

$$|d - 8| = 45 \qquad\qquad |d - 8| = 45$$
$$d - 8 = 45 \qquad\qquad d - 8 = -45$$
$$d - 8 + 8 = 45 + 8 \qquad d - 8 + 8 = -45 + 8$$
$$d = 45 + 8 \qquad\qquad d = -45 + 8$$
$$d = 53 \qquad\qquad d = -37$$

12. The solutions are $a = 2$ or $a = -4$.

$$|a + 1| = 3 \qquad\qquad |a + 1| = 3$$
$$a + 1 = 3 \qquad\qquad a + 1 = -3$$
$$a + 1 - 1 = 3 - 1 \qquad a + 1 - 1 = -3 - 1$$
$$a = 3 - 1 \qquad\qquad a = -3 - 1$$
$$a = 2 \qquad\qquad a = -4$$

13. The solutions are $j = 19$ or $j = -1$.

$$|j - 9| = 10 \qquad\qquad |j - 9| = 10$$
$$j - 9 = 10 \qquad\qquad j - 9 = -10$$
$$j - 9 + 9 = 10 + 9 \qquad j - 9 + 9 = -10 + 9$$
$$j = 10 + 9 \qquad\qquad j = -10 + 9$$
$$j = 19 \qquad\qquad j = -1$$

14. The solutions are $s = 2$ or $s = -12$.

$$|s + 5| = 7 \qquad\qquad |s + 5| = 7$$
$$s + 5 = 7 \qquad\qquad s + 5 = -7$$
$$s + 5 - 5 = 7 - 5 \qquad s + 5 - 5 = -7 - 5$$
$$s = 7 - 5 \qquad\qquad s = -7 - 5$$
$$s = 2 \qquad\qquad s = -12$$

15. The solutions are $f = 12$ or $f = -4$.

$$|f - 4| = 8 \qquad\qquad |f - 4| = 8$$
$$f - 4 = 8 \qquad\qquad f - 4 = -8$$
$$f - 4 + 4 = 8 + 4 \qquad f - 4 + 4 = -8 + 4$$
$$f = 8 + 4 \qquad\qquad f = -8 + 4$$
$$f = 12 \qquad\qquad f = -4$$

16. The solutions are $k = 2$ or $k = -\dfrac{6}{7}$.

$$|7k - 4| = 10 \qquad\qquad |7k - 4| = 10$$
$$7k - 4 = 10 \qquad\qquad 7k - 4 = -10$$
$$7k - 4 + 4 = 10 + 4 \qquad 7k - 4 + 4 = -10 + 4$$
$$7k = 14 \qquad\qquad 7k = -6$$
$$\frac{7k}{7} = \frac{14}{7} \qquad\qquad \frac{7k}{7} = \frac{-6}{7}$$
$$k = 2 \qquad\qquad k = -\frac{6}{7}$$

17. The solutions are $d = 8$ or $d = -16$.

$$|2d + 8| = 24 \qquad\qquad |2d + 8| = 24$$
$$2d + 8 = 24 \qquad\qquad 2d + 8 = -24$$
$$2d + 8 - 8 = 24 - 8 \qquad 2d + 8 - 8 = -24 - 8$$
$$2d = 16 \qquad\qquad 2d = -32$$
$$\frac{2d}{2} = \frac{16}{2} \qquad\qquad \frac{2d}{2} = \frac{-32}{2}$$
$$d = 8 \qquad\qquad d = -16$$

18. The solutions are $t = 100$ or $t = -100$.

$$|t \div 5| = 20 \qquad\qquad |t \div 5| = 20$$
$$t \div 5 = 20 \qquad\qquad t \div 5 = -20$$
$$\frac{t}{5} = 20 \qquad\qquad \frac{t}{5} = -20$$
$$5 \times \left(\frac{t}{5}\right) = 20 \times 5 \qquad 5 \times \left(\frac{t}{5}\right) = -20 \times 5$$
$$t = 20 \times 5 \qquad\qquad t = -20 \times 5$$
$$t = 100 \qquad\qquad t = -100$$

19. The solutions are $y = 7$ or $y = -3$.

$$|2y - 4| = 10 \qquad\qquad |2y - 4| = 10$$
$$2y - 4 = 10 \qquad\qquad 2y - 4 = -10$$
$$2y - 4 + 4 = 10 + 4 \qquad 2y - 4 + 4 = -10 + 4$$
$$2y = 14 \qquad\qquad 2y = -6$$
$$\frac{2y}{2} = \frac{14}{2} \qquad\qquad \frac{2y}{2} = \frac{-6}{2}$$
$$y = 7 \qquad\qquad y = -3$$

20. The solutions are $x = 18$ or $x = -18$.

$$|3x \div 3| = 18 \qquad\qquad |3x \div 3| = 18$$
$$3x \div 3 = 18 \qquad\qquad 3x \div 3 = -18$$
$$\frac{3x}{3} = 18 \qquad\qquad \frac{3x}{3} = -18$$
$$3 \times \left(\frac{3x}{3}\right) = 18 \times 3 \qquad 3 \times \left(\frac{3x}{3}\right) = -18 \times 3$$
$$3x = 54 \qquad\qquad 3x = -54$$
$$\frac{3x}{3} = \frac{54}{3} \qquad\qquad \frac{3x}{3} = \frac{-54}{3}$$
$$x = 18 \qquad\qquad x = -18$$

21. The solutions are $m = \dfrac{1}{2}$ or $m = -\dfrac{1}{2}$.

$$|4m \times 8| = 16 \qquad\qquad |4m \times 8| = 16$$
$$4m \times 8 = 16 \qquad\qquad 4m \times 8 = -16$$
$$8(4m) = 16 \qquad\qquad 8(4m) = -16$$
$$\frac{8(4m)}{8} = \frac{16}{8} \qquad\qquad \frac{8(4m)}{8} = \frac{-16}{8}$$
$$4m = 2 \qquad\qquad 4m = -2$$
$$\frac{4m}{4} = \frac{2}{4} \qquad\qquad \frac{4m}{4} = \frac{-2}{4}$$
$$m = \frac{1}{2} \qquad\qquad m = -\frac{1}{2}$$

22. The solutions are $e = 112$ or $e = -112$.

$$|e \div 4| = 28 \qquad\qquad |e \div 4| = 28$$
$$\frac{e}{4} = 28 \qquad\qquad \frac{e}{4} = -28$$
$$4 \times \left(\frac{e}{4}\right) = 28 \times 4 \qquad 4 \times \left(\frac{e}{4}\right) = -28 \times 4$$
$$e = 28 \times 4 \qquad\qquad e = -28 \times 4$$
$$e = 112 \qquad\qquad e = -112$$

23. The solutions are $u = 9$ or $u = -9$.

$$|6u \div 18| = 3 \qquad\qquad |6u \div 18| = 3$$

$$6u \div 18 = 3 \qquad\qquad 6u \div 18 = -3$$

$$\frac{6u}{18} = 3 \qquad\qquad \frac{6u}{18} = -3$$

$$18 \times \left(\frac{6u}{18}\right) = 3 \times 18 \qquad 18 \times \left(\frac{6u}{18}\right) = -3 \times 18$$

$$6u = 54 \qquad\qquad 6u = -54$$

$$\frac{6u}{6} = \frac{54}{6} \qquad\qquad \frac{6u}{6} = \frac{-54}{6}$$

$$u = 9 \qquad\qquad u = -9$$

25. The solutions are $v = 10$ or $v = -12.5$.

$$|4v + 5| = 45 \qquad\qquad |4v + 5| = 45$$

$$4v + 5 = 45 \qquad\qquad 4v + 5 = -45$$

$$4v + 5 - 5 = 45 - 5 \qquad 4v + 5 - 5 = -45 - 5$$

$$4v = 40 \qquad\qquad 4v = -50$$

$$\frac{4v}{4} = \frac{40}{4} \qquad\qquad \frac{4v}{4} = \frac{-50}{4}$$

$$v = 10 \qquad\qquad v = -12.5$$

24. The solutions are $c = \dfrac{5}{9}$ or $c = -\dfrac{5}{9}$.

$$|9c \times 2| = 10 \qquad\qquad |9c \times 2| = 10$$

$$9c \times 2 = 10 \qquad\qquad 9c \times 2 = -10$$

$$2(9c) = 10 \qquad\qquad 2(9c) = -10$$

$$\frac{2(9c)}{2} = \frac{10}{2} \qquad\qquad \frac{2(9c)}{2} = \frac{-10}{2}$$

$$9c = 5 \qquad\qquad 9c = -5$$

$$\frac{9c}{9} = \frac{5}{9} \qquad\qquad \frac{9c}{9} = \frac{-5}{9}$$

$$c = \frac{5}{9} \qquad\qquad c = -\frac{5}{9}$$

Chapter 9

Inequalities

In this chapter, we'll review
the following concepts:

What is an inequality?
Parts of inequalities
Writing inequalities
Solving inequalities

What is an inequality?

An inequality is similar to an equation. Like an equation, it is a true mathematical statement. Unlike equations, inequalities do not contain traditional equal signs (=).

Parts of inequalities

Algebra inequalities contain expressions and sometimes a single number or variable. Instead of containing equal signs, they contain one of the four following symbols:

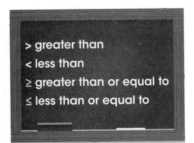

> greater than
< less than
≥ greater than or equal to
≤ less than or equal to

Here are some examples of inequalities using variables:

$$x > 1$$
$$y < 9$$
$$r \geq 7$$
$$p \leq 15$$

The first statement tells us that x is greater than 1. The value of x can be any value larger than 1, such as 2 or 100. The second inequality tells us that y is less than 9. The value of y can be any value less than 9, such as 3 or 0.5.

egghead's Guide to Algebra

The third statement contains a "greater than or equal to" sign. This tells us that r is greater than or equal to 7. The value of r could be 7, or it could also be a larger number. The value of r is greater than 7 and also includes 7.

The fourth inequality describes a variable, p, that is less than or equal to 15. The value of p could be 15 or any number less than 15, such as 10 or $3\frac{1}{3}$.

Writing inequalities

To work with inequalities, like with equations, we may need to translate English descriptions into math. There are a number of ways to describe different inequality relationships. Here are some examples.

To indicate a quantity greater than 4, we might see the phrase "x is greater than 4." That phrase would be written as follows:

$$x > 4$$

To describe a quantity that includes 4, we might see "x is greater than or equal to 4."

$$x \geq 4$$

This relationship might also be described as "x is at least 4" or "x is a minimum of 4."

All three phrases mean the same thing.

To write the phrase "y is a maximum of 10," we'd use the $\leq$ sign:

$$y \leq 10$$

This means that the value of *y* is less than or equal to 10.

Try some translations now for practice.

Practice Questions—Writing inequalities

Directions: Write the following inequalities in symbol form. You will find the Practice Question Solutions on page 194.

1. *b* is greater than 8

2. *y* is less than or equal to 60

3. *w* is at most 20

4. *c* is less than 40

5.

6. k^2 is no more than 16

7. *r* is a maximum of 32

8. z^3 is a minimum of 3

9. *p* is not equal to 7

10. f^3 is at least 4

m is at least 25

Solving inequalities

To solve inequalities, we follow a similar process as we do when solving equations. First, simplify the inequality. Then isolate the variable on one side of the inequality sign.

$$7 + c \leq 14$$

To solve this inequality, we would simplify by moving the 7 to the right side of the $\leq$ sign. Subtract 7 from both sides of the inequality:

$$7 + c \leq 14$$
$$7 - 7 + c \leq 14 - 7$$
$$c \leq 14 - 7$$
$$c \leq 7$$

This leaves c isolated on the left side of the inequality, and we have our answer: $c \leq 7$.

In this case, just simplifying by itself solved the problem.

More complex inequalities require more steps to solve.

Examples

In this inequality, we must use division to undo multiplication:

$$6z + 4 > 34$$

As you would with an equation, first subtract the 4 from both sides:

$$6z + 4 > 34$$
$$6z + 4 - 4 > 34 - 4$$
$$6z > 34 - 4$$
$$6z > 30$$

Now divide by 6 to isolate the variable:

$$6z > 30$$
$$\frac{6z}{6} > \frac{30}{6}$$
$$z > \frac{30}{6}$$
$$z > 5$$

The correct answer is $z > 5$.

When solving inequalities, there is one major difference from solving equations. If we multiply or divide an inequality by a negative number, we must change the direction of the inequality sign.

$$-12a \leq 60$$

To isolate the a on the left, we must divide both sides of the inequality by -12. This requires flipping the $\leq$ sign to the other direction:

$$-12a \leq 60$$
$$\frac{-12a}{-12} \leq \frac{60}{-12}$$
$$a \geq \frac{60}{-12}$$
$$a \geq -5$$

The $\leq$ becomes $\geq$!

Flipping the inequality sign is important when you are multiplying by a negative number, too:

$$5r \div -10 > 20$$
$$\frac{5r}{-10} > 20$$
$$-10\left(\frac{5r}{-10}\right) > 20 \times -10$$
$$5r < 20 \times -10$$
$$5r < -200$$
$$\frac{5r}{5} < \frac{-200}{5}$$
$$r < \frac{-200}{5}$$
$$r < -40$$

In this inequality, we multiplied both sides by -10 and flipped the inequality sign.

Practice Questions—Solving inequalities

Directions: Solve the inequalities shown. You will find the Practice Question Solutions on page 194.

11. $x + 3 > 5$

12. $a - 7 < 12$

13. $j + 10 < 50$

14. $w - 2 > 9$

15. $p + 6 > 24$

16. $2d + 7 > 12$

17. $-14y < 3$

HINT: Remember to flip the inequality sign!

18. $5a - 10 > 25$

19. $4r \div 2 < 20$

20. $-8z < 32$

Chapter Review

Directions: Perform the operations shown. Solutions can be found on page 196.

1. Write the symbolic inequality statement for "*b* is greater than 2 + 5."

2. Write the symbolic inequality statement for "*x* is at most 59."

3. Write the symbolic inequality statement for "*r* is not equal to 5 + 3."

4. Write the symbolic inequality statement for "*v* is a maximum of 7."

5. Write the symbolic inequality statement for "*q* is less than 10 – 4."

6. Write the symbolic inequality statement for "*m* is at least 91."

7. Write the symbolic inequality statement for "*t* is a minimum of 20 – 2."

8. Write the symbolic inequality statement for "h^2 is no more than 9."

9. Write the symbolic inequality statement for "d^3 is no less than 31."

10. If $e + 5 > 14$, what is the value of e?

11. Solve for w, if $w - 4 < 16$.

12. If $g + 50 > 90$, what is the value of g?

13. If $-10y < 2$, what is the value of y?

14. Solve for a, if $4a - 8 > 40$.

15. If $2r \div 4 < 60$, what is the value of r?

16. Solve for z, if $-7z > 28$.

17. Solve for j, if $j + 1 < 5$.

18. If $8p \div 2 < 16$, what is the value of p?

19. Solve for s, if $3s + 6 > 18$.

20. Solve for c, if $5c - 7 > 12$.

21. What is the value of x if $3(x + 4) \geq 4x + 2$?

22. If $\frac{1}{2}a + 7 < -2$, find the value of a.

23. What is the value of k if $2(k + 4) \geq 42$?

24. If $\frac{1}{3}n - 9 > -21$, find the value of n.

25. What is the value of y if $9(y - 3) \geq 2y + 12$?

4. $c < 40$

5. $m \geq 25$

6. $k^2 \leq 16$

7. $r \leq 32$

8. $z^3 \geq 3$

9. $p \neq 7$

10. $f^3 \geq 4$

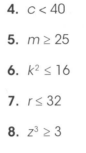

Nice work!

Practice Question Solutions

Writing inequalities

1. $b > 8$

2. $y \leq 60$

3. $w \leq 20$

Solving inequalities

11. The correct answer is $x > 2$.

$$x + 3 > 5$$
$$x + 3 - 3 > 5 - 3$$
$$x > 5 - 3$$
$$x > 2$$

12. The correct answer is $a < 19$.

$$a - 7 < 12$$
$$a - 7 + 7 < 12 + 7$$
$$a < 19$$

egghead's Guide to Algebra

13. The correct answer is $j < 40$.

$$j + 10 < 50$$
$$j + 10 - 10 < 50 - 10$$
$$j < 40$$

14. The correct answer is $w > 11$.

$$w - 2 > 9$$
$$w - 2 + 2 > 9 + 2$$
$$w > 11$$

15. The correct answer is $p > 18$.

$$p + 6 > 24$$
$$p + 6 - 6 > 24 - 6$$
$$p > 24 - 6$$
$$p > 18$$

16. The correct answer is $d > 2.5$.

$$2d + 7 > 12$$
$$2d + 7 - 7 > 12 - 7$$
$$2d > 5$$
$$\frac{2d}{2} > \frac{5}{2}$$
$$d > 2.5$$

This could also be left in fractional form and written as $d > \frac{5}{2}$.

17. The correct answer is $y > -\frac{3}{14}$.

$$-14y < 3$$
$$\frac{-14y}{-14} < \frac{3}{-14}$$
$$y > -\frac{3}{14}$$

18. The correct answer is $a > 7$.

$$5a - 10 > 25$$
$$5a - 10 + 10 > 25 + 10$$
$$5a > 35$$
$$\frac{5a}{5} > \frac{35}{5}$$
$$a > \frac{35}{5}$$
$$a > 7$$

19. The correct answer is $r < 10$.

$$4r \div 2 < 20$$
$$\frac{4r}{2} < 20$$
$$2 \times \left(\frac{4r}{2}\right) < 20 \times 2$$
$$4r < 40$$
$$\frac{4r}{4} < \frac{40}{4}$$
$$r < \frac{40}{4}$$
$$r < 10$$

20. The correct answer is $z > -4$.

$$-8z < 32$$
$$\frac{-8z}{-8} < \frac{32}{-8}$$
$$z > \frac{32}{-8}$$
$$z > -4$$

Chapter Review Solutions

1. $b > 2 + 5$

2. $x \le 59$

3. $r \ne 5 + 3$

4. $v \le 7$

5. $q < 10 - 4$

6. $m \ge 91$

7. $t \ge 20 - 2$

8. $h^2 \le 9$

9. $d^3 \ge 31$

10. The correct answer is $e > 9$.

$$e + 5 > 14$$
$$e + 5 - 5 > 14 - 5$$
$$e > 14 - 5$$
$$e > 9$$

11. The correct answer is $w < 20$.

$$w - 4 < 16$$
$$w - 4 + 4 < 16 + 4$$
$$w < 16 + 4$$
$$w < 20$$

12. The correct answer is $g > 40$.

$$g + 50 > 90$$
$$g + 50 - 50 > 90 - 50$$
$$g > 90 - 50$$
$$g > 40$$

13. The correct answer is $y > -\frac{1}{5}$.

$$-10y < 2$$
$$\frac{-10y}{-10} < \frac{2}{-10}$$
$$y > \frac{2}{-10}$$
$$y > -\frac{1}{5}$$

14. The correct answer is $a > 12$.

$$4a - 8 > 40$$
$$4a - 8 + 8 > 40 + 8$$
$$4a > 48$$
$$\frac{4a}{4} > \frac{48}{4}$$
$$a > \frac{48}{4}$$
$$a > 12$$

15. The correct answer is $r < 120$.

$$2r \div 4 < 60$$
$$\frac{2r}{4} < 60$$
$$4 \times \left(\frac{2r}{4}\right) < 60 \times 4$$
$$2r < 240$$
$$\frac{2r}{2} < \frac{240}{2}$$
$$r < \frac{240}{2}$$
$$r < 120$$

Chapter 9: Inequalities

16. The correct answer is $z < -4$.

$$-7z > 28$$
$$\frac{-7z}{-7} > \frac{28}{-7}$$
$$z < \frac{28}{-7}$$
$$z < -4$$

17. The correct answer is $j < 4$.

$$j + 1 < 5$$
$$j + 1 - 1 < 5 - 1$$
$$j < 5 - 1$$
$$j < 4$$

18. The correct answer is $p < 4$.

$$8p \div 2 < 16$$
$$\frac{8p}{2} < 16$$
$$2 \times \left(\frac{8p}{2}\right) < 16 \times 2$$
$$8p < 32$$
$$\frac{8p}{8} < \frac{32}{8}$$
$$p < \frac{32}{8}$$
$$p < 4$$

19. The correct answer is $s > 4$.

$$3s + 6 > 18$$
$$3s + 6 - 6 > 18 - 6$$
$$3s > 12$$
$$\frac{3s}{3} > \frac{12}{3}$$
$$s > \frac{12}{3}$$
$$s > 4$$

20. The correct answer is $c > 3.8$.

$$5c - 7 > 12$$
$$5c - 7 + 7 > 12 + 7$$
$$5c > 19$$
$$\frac{5c}{5} > \frac{19}{5}$$
$$c > \frac{19}{5}$$
$$c > 3.8$$

This could also be kept in fractional form and written as $c > \frac{19}{5}$.

21. The correct answer is $x \leq 10$.

$$3(x + 4) \geq 4x + 2$$
$$3x + 12 \geq 4x + 2$$
$$3x - 4x + 12 \geq 4x - 4x + 2$$
$$-x + 12 \geq 4x - 4x + 2$$
$$-x + 12 \geq 2$$
$$-x + 12 - 12 \geq 2 - 12$$
$$-x \geq 2 - 12$$
$$-x \geq -10$$
$$\frac{-1x}{-1} \geq \frac{-10}{-1}$$
$$x \leq \frac{-10}{-1}$$
$$x \leq 10$$

22. The correct answer is $a < -18$.

$$\frac{1}{2}a + 7 < -2$$
$$\frac{1}{2}a + 7 - 7 < -2 - 7$$
$$\frac{1}{2}a < -2 - 7$$
$$\frac{1}{2}a < -9$$
$$2 \times \left(\frac{1}{2}a\right) < -9 \times 2$$
$$a < -9 \times 2$$
$$a < -18$$

23. The correct answer is $k \geq 17$.

$$2(k + 4) \geq 42$$
$$2k + 8 \geq 42$$
$$2k + 8 - 8 \geq 42 - 8$$
$$2k \geq 42 - 8$$
$$2k \geq 34$$
$$\frac{2k}{2} \geq \frac{34}{2}$$
$$k \geq \frac{34}{2}$$
$$k \geq 17$$

24. The correct answer is $n > -36$.

$$\frac{1}{3}n - 9 > -21$$
$$\frac{1}{3}n - 9 + 9 > -21 + 9$$
$$\frac{1}{3}n > -21 + 9$$
$$\frac{1}{3}n > -12$$
$$3 \times \left(\frac{1}{3}n\right) > -12 \times 3$$
$$n > -12 \times 3$$
$$n > -36$$

25. The correct answer is $y \geq 5.57$.

$$9(y - 3) \geq 2y + 12$$
$$9y - 27 \geq 2y + 12$$
$$9y - 2y - 27 \geq 2y - 2y + 12$$
$$7y - 27 \geq 2y - 2y + 12$$
$$7y - 27 \geq 12$$
$$7y - 27 + 27 \geq 12 + 27$$
$$7y \geq 12 + 27$$
$$7y \geq 39$$
$$\frac{7y}{7} \geq \frac{39}{7}$$
$$y \geq \frac{39}{7}$$
$$y \geq 5.57$$

This could also be kept in fractional form and written as $y \geq \frac{39}{7}$.

Chapter 10

Systems of Equations

In this chapter, we'll review the following concepts:

What are systems of equations?
Solving systems of equations by substitution
Solving systems of equations by combining

What are systems of equations?

Systems of equations are sets consisting of two or more equations. Each equation in the set contains one or more variables.

This system of equations has two variables, x and y.

$$\begin{cases} 3x + y = 12 \\ 2x - 4y = 17 \end{cases}$$

This system of equations has three equations and three variables:

$$\begin{cases} 2x + 3y + z = 4 \\ x + y + z = 9 \\ 3x + 4y + 2z = 11 \end{cases}$$

In this chapter, we will work with systems of equations containing two variables and no exponents.

Systems of equations are also called **simultaneous equations**.

egghead's Guide to Algebra

Solving systems of equations by substitution

To solve a system of equations involving two variables, we must find a single value for each variable that satisfies both equations.

$$\begin{cases} x + y = 2 \\ 3x - 2y = 1 \end{cases}$$

We are looking for one value for x and one value for y that work in both equations. When solving systems of equations, we write the solution in parentheses, like this:

$$(x, y)$$

If the value of x is 5 and the value of y is 3, the solution would be written (5, 3).

There are several ways to solve systems of equations. Let's start with the substitution method. To solve a system of equations using substitution, we first find the value of one variable in terms of the other. In the example shown, the first equation is the easiest to work with.

$$\begin{cases} x + y = 2 \\ 3x - 2y = 1 \end{cases}$$

Solve the first equation for y. We call this "solving for y in terms of x." Subtract x from both sides of the equation:

$$x + y = 2$$
$$x - x + y = 2 - x$$
$$y = 2 - x$$

This gives us the value $y = 2 - x$.

Substitute $2 - x$ for y in the second equation:

$$3x - 2y = 1$$
$$3x - 2(2 - x) = 1$$
$$3x - 4 + 2x = 1$$
$$3x + 2x - 4 = 1$$
$$5x - 4 = 1$$
$$5x - 4 + 4 = 1 + 4$$
$$5x = 1 + 4$$
$$5x = 5$$
$$\frac{5x}{5} = \frac{5}{5}$$
$$x = \frac{5}{5}$$
$$x = 1$$

The value of x is 1. Now, substitute 1 for x in either equation and solve. We'll use the first equation:

$$x + y = 2$$
$$(1) + y = 2$$
$$1 - 1 + y = 2 - 1$$
$$y = 2 - 1$$
$$y = 1$$

The value of y is also 1. The solution to the system of equations is $(1, 1)$.

Practice Questions—Solving systems of equations by substitution

Directions: Perform the operations shown. You will find the Practice Question Solutions on page 213.

1. Solve the system of equations using substitution: $\begin{cases} 3x + y = 2 \\ 4x + 2y = 4 \end{cases}$

2. Solve the system of equations using substitution: $\begin{cases} 3y - 2x = -1 \\ y + 2x = 5 \end{cases}$

3. Solve the system of equations using substitution: $\begin{cases} -x + y = 1 \\ 2x + y = 10 \end{cases}$

4. Solve the system of equations using substitution: $\begin{cases} 4x - y = 10 \\ -x + y = 5 \end{cases}$

5. Solve the system of equations using substitution: $\begin{cases} 2x + 2y = 0 \\ 6x + y = -10 \end{cases}$

6. Solve the system of equations using substitution: $\begin{cases} 2x + y = -4 \\ 3x + y = 9 \end{cases}$

7. Solve the system of equations using substitution: $\begin{cases} 3x - y = 3 \\ 4x - 8y = 36 \end{cases}$

8. Solve the system of equations using

substitution: $\begin{cases} 6y - 2x = 6 \\ 3y - 4x = -6 \end{cases}$

10. Solve the system of equations using

substitution: $\begin{cases} -3x + y = -1 \\ -2x + y = 2 \end{cases}$

9. Solve the system of equations using

substitution: $\begin{cases} 5y + 4x = 48 \\ y - 3x = 2 \end{cases}$

egghead's Guide to Algebra

Solving systems of equations by combining

In addition to using substitution, we can also solve systems of equations by combining. The combining method works best when there are variables in the equations that cancel each other out.

$$\begin{cases} 2x + 3y = 13 \\ 4x - 3y = -1 \end{cases}$$

To combine two equations, we add or subtract them to cause one of the variables to drop out. We can choose to either add or subtract the equations, based on which operation will eliminate one of the variables. In the example shown, one of the equations has the term $+3y$, and the other equation has the term $-3y$. If we add the two equations, the y variables will cancel each other out:

$$\begin{array}{r} 2x + 3y = 13 \\ + \ 4x - 3y = -1 \\ \hline 6x + 0y = 12 \end{array}$$

Here, we added the first two terms to produce $2x + 4x = 6x$. The second two terms, the y terms, cancelled each other out: $3y - 3y = 0y$, or 0. The terms on the right side of the equal signs add up to $13 - 1$, or 12.

This gives us a new equation to work with: $6x + 0y = 12$. Use this equation to solve for x:

$$6x + 0y = 12$$
$$6x = 12$$
$$\frac{6x}{6} = \frac{12}{6}$$
$$x = \frac{12}{6}$$
$$x = 2$$

The value of x is 2. Substitute 2 for x in either of the original equations and solve:

$$2x + 3y = 13$$
$$2(2) + 3y = 13$$
$$4 + 3y = 13$$
$$4 - 4 + 3y = 13 - 4$$
$$3y = 13 - 4$$
$$3y = 9$$
$$\frac{3y}{3} = \frac{9}{3}$$
$$y = \frac{9}{3}$$
$$y = 3$$

The value of y is 3. The solution of the system of equations is $(2, 3)$.

Multiplying before combining

At times, you will have to multiply one or both of the equations before combining them. Use multiplication to create variables that cancel each other out.

$$\begin{cases} 3x + 2y = 8 \\ 5x + 4y = 14 \end{cases}$$

In this system of equations, there are no obvious terms that would cancel each other out through addition or subtraction. However, we can multiply one or both of the equations by a number to create terms that will cancel out when the equations are combined.

Multiply all terms of the first equation by 2:

$$3x + 2y = 8$$
$$\underline{\times 2}$$
$$6x + 4y = 16$$

This creates a new equation, $6x + 4y = 16$. Now we have a term, $4y$, that will cancel out the $4y$ in the second equation.

Equations can be combined by either addition or subtraction. So, subtract the second equation from the first:

$$6x + 4y = 16$$
$$\underline{- \ 5x + 4y = 14}$$
$$x + 0y = 2$$

This gives us an easier equation to work with. Solve for x:

$$x + 0y = 2$$
$$x = 2$$

The value of x is 2. Substitute 2 for x in either of the original equations and solve:

$$3x + 2y = 8$$
$$3(2) + 2y = 8$$
$$6 + 2y = 8$$
$$6 - 6 + 2y = 8 - 6$$
$$2y = 8 - 6$$
$$2y = 2$$
$$\frac{2y}{2} = \frac{2}{2}$$
$$y = \frac{2}{2}$$
$$y = 1$$

The value of y is 1. The solution of the system of equations is $(2, 1)$.

You can use multiplication if necessary to make a variable cancel out when equations are combined.

One or both equations may need to be multiplied before combining.

Practice Questions—Solving systems of equations by combining

Directions: Perform the operations shown. You will find the Practice Question Solutions on page 218.

11. Solve the system of equations by combining: $\begin{cases} 5x + 6y = 11 \\ 3x + 6y = 9 \end{cases}$

14. Solve the system of equations by combining: $\begin{cases} 4x - y = 5 \\ 2x - 4y = 6 \end{cases}$

12. Solve the system of equations by combining: $\begin{cases} 7x + 2y = 14 \\ 6x - 2y = 12 \end{cases}$

15. Solve the system of equations by combining: $\begin{cases} x - y = 12 \\ x + y = 8 \end{cases}$

13. Solve the system of equations by combining: $\begin{cases} 4x + 3y = 10 \\ 2x + 3y = 14 \end{cases}$

16. Solve the system of equations by combining: $\begin{cases} -5x + y = 1 \\ 9x - 3y = 0 \end{cases}$

17. Solve the system of equations by combining: $\begin{cases} 3x - 2y = 17 \\ -2x - 5y = -5 \end{cases}$

18. Solve the system of equations by

combining: $\begin{cases} -2x + y = 2 \\ -4x + 4y = 4 \end{cases}$

20. Solve the system of equations by

combining: $\begin{cases} 3x + y = 14 \\ x - y = -2 \end{cases}$

19. Solve the system of equations by

combining: $\begin{cases} 3x + 3y = 0 \\ 4x - y = -1 \end{cases}$

Chapter Review

Directions: Perform the operations shown. Solutions can be found on page 223.

1. Solve the system of equations using

   substitution: $\begin{cases} 5x + y = -12 \\ x + 2y = 3 \end{cases}$

2. Solve the system of equations using

   substitution: $\begin{cases} 2x - 4y = -20 \\ 3x - y = 0 \end{cases}$

3. Solve the system of equations using

   substitution: $\begin{cases} x + 3y = 2 \\ -2x + 4y = 6 \end{cases}$

4. Solve the system of equations using

   substitution: $\begin{cases} 2y - 3x = -2 \\ y - 2x = 1 \end{cases}$

5. Solve the system of equations using

   substitution: $\begin{cases} y - x = 1 \\ 4y - 8x = 8 \end{cases}$

6. Solve the system of equations using

   substitution: $\begin{cases} 5x - y = 1 \\ 3x - 2y = -12 \end{cases}$

7. Solve the system of equations using

   substitution: $\begin{cases} x - 3y = -1 \\ x - y = -5 \end{cases}$

8. Solve the system of equations using

substitution: $\begin{cases} x - 3y = 2 \\ 2x + 6y = 52 \end{cases}$

13. Solve the system of equations using

substitution: $\begin{cases} 3y - 6x = 12 \\ 2y - 3x = -2 \end{cases}$

9. Solve the system of equations using

substitution: $\begin{cases} y - x = 2 \\ 2y - 3x = 0 \end{cases}$

14. Solve the system of equations by

combining: $\begin{cases} x + y = 1 \\ x + 5y = 21 \end{cases}$

10. Solve the system of equations using

substitution: $\begin{cases} 2x - 3y = 0 \\ 4x - 12y = 24 \end{cases}$

15. Solve the system of equations by

combining: $\begin{cases} 3x + y = 13 \\ x + 6y = -7 \end{cases}$

11. Solve the system of equations using

substitution: $\begin{cases} -x + y = 5 \\ x - 2y = 2 \end{cases}$

16. Solve the system of equations by

combining: $\begin{cases} 5x - y = 3 \\ -10x + 5y = 0 \end{cases}$

12. Solve the system of equations using

substitution: $\begin{cases} y - x = 2 \\ 3x + 5y = 10 \end{cases}$

17. Solve the system of equations by

combining: $\begin{cases} x + 2y = 8 \\ x - 2y = 4 \end{cases}$

22. Solve the system of equations by

combining: $\begin{cases} 2x + y = 9 \\ 6x - 2y = 32 \end{cases}$

18. Solve the system of equations by

combining: $\begin{cases} 4x + 7y = 14 \\ 4x - 5y = -10 \end{cases}$

23. Solve the system of equations by

combining: $\begin{cases} 2x - y = 12 \\ 3x + 4y = 18 \end{cases}$

19. Solve the system of equations by

combining: $\begin{cases} 2x - 3y = -2 \\ 4x + y = 24 \end{cases}$

24. Solve the system of equations by

combining: $\begin{cases} 4x - 3y = 25 \\ -3x + 8y = 10 \end{cases}$

20. Solve the system of equations by

combining: $\begin{cases} 9x + y = 36 \\ x + y = 12 \end{cases}$

25. Solve the system of equations by

combining: $\begin{cases} x - 2y = 6 \\ 4x - 2y = 12 \end{cases}$

21. Solve the system of equations by

combining: $\begin{cases} 7x + 2y = 13 \\ x + y = 24 \end{cases}$

**Practice
Question
Solutions**

Solving systems of equations by substitution

1. The correct answer is (0, 2).

Solve the first equation for y:

$$3x + y = 2$$
$$3x - 3x + y = 2 - 3x$$
$$y = 2 - 3x$$

Substitute $2 - 3x$ for y in the second equation:

$$4x + 2y = 4$$
$$4x + 2(2 - 3x) = 4$$
$$4x + 4 - 6x = 4$$
$$4x - 6x + 4 = 4$$
$$-2x + 4 = 4$$
$$-2x + 4 - 4 = 4 - 4$$
$$-2x = 4 - 4$$
$$-2x = 0$$
$$x = 0$$

Substitute 0 for x in either equation:

$$3x + y = 2$$
$$3(0) + y = 2$$
$$0 + y = 2$$
$$y = 2$$

2. The correct answer is (2, 1).

Solve the second equation for y:

$$y + 2x = 5$$
$$y + 2x - 2x = 5 - 2x$$
$$y = 5 - 2x$$

Substitute $5 - 2x$ for y in the first equation:

$$3y - 2x = -1$$
$$3(5 - 2x) - 2x = -1$$
$$15 - 6x - 2x = -1$$
$$15 - 8x = -1$$
$$15 - 15 - 8x = -1 - 15$$
$$-8x = -1 - 15$$
$$-8x = -16$$
$$\frac{-8x}{-8} = \frac{-16}{-8}$$
$$x = \frac{-16}{-8}$$
$$x = 2$$

Substitute 2 for x in either equation:

$$3y - 2x = -1$$
$$3y - 2(2) = -1$$
$$3y - 4 = -1$$
$$3y - 4 + 4 = -1 + 4$$
$$3y = 3$$
$$\frac{3}{3}y = \frac{3}{3}$$
$$y = \frac{3}{3}$$
$$y = 1$$

3.

The correct answer is $(3, 4)$.

Solve the first equation for y:

$$-x + y = 1$$
$$x - x + y = 1 + x$$
$$y = 1 + x$$

Substitute $1 + x$ for y in the second equation:

$$2x + y = 10$$
$$2x + (1 + x) = 10$$
$$2x + 1 + x = 10$$
$$3x + 1 = 10$$
$$3x + 1 - 1 = 10 - 1$$
$$3x = 10 - 1$$
$$3x = 9$$
$$\frac{3x}{3} = \frac{9}{3}$$
$$x = \frac{9}{3}$$
$$x = 3$$

Substitute 3 for x in either equation:

$$-x + y = 1$$
$$-(3) + y = 1$$
$$-3 + 3 + y = 1 + 3$$
$$y = 1 + 3$$
$$y = 4$$

4. The correct answer is $(5, 10)$.

Solve the second equation for y:

$$-x + y = 5$$
$$x - x + y = 5 + x$$
$$y - 5 + x$$

Substitute $5 + x$ for y in the first equation:

$$4x - y = 10$$
$$4x - (5 + x) = 10$$
$$4x - 5 - x = 10$$
$$3x - 5 = 10$$
$$3x - 5 + 5 = 10 + 5$$
$$3x = 10 + 5$$
$$3x = 15$$
$$\frac{3x}{3} = \frac{15}{3}$$
$$x = \frac{15}{3}$$
$$x = 5$$

Substitute 5 for x in either equation:

$$4x - y = 10$$
$$4(5) - y = 10$$
$$20 - y = 10$$
$$20 - 20 - y = 10 - 20$$
$$-y = 10 - 20$$
$$-y = -10$$
$$\frac{-y}{-1} = \frac{-10}{-1}$$
$$y = \frac{-10}{-1}$$
$$y = 10$$

5. The correct answer is $(-2, 2)$.

Solve the second equation for y:

$$6x + y = -10$$
$$6x - 6x + y = -10 - 6x$$
$$y = -10 - 6x$$

Substitute $-10 - 6x$ for y in the first equation:

$$2x + 2y = 0$$
$$2x + 2(-10 - 6x) = 0$$
$$2x - 20 - 12x = 0$$
$$-10x - 20 = 0$$
$$-10x - 20 + 20 = 0 + 20$$
$$-10x = 0 + 20$$
$$-10x = 20$$
$$\frac{-10x}{-10} = \frac{20}{-10}$$
$$x = \frac{20}{-10}$$
$$x = -2$$

Substitute -2 for x in either equation:

$$6x + y = -10$$
$$6(-2) + y = -10$$
$$-12 + y = -10$$
$$-12 + 12 + y = -10 + 12$$
$$y = -10 + 12$$
$$y = 2$$

6. The correct answer is $(13, -30)$.

Solve the first equation for y:

$$2x + y = -4$$
$$2x - 2x + y = -4 - 2x$$
$$y = -4 - 2x$$

Substitute $-4 - 2x$ for y in the second equation:

$$3x + y = 9$$
$$3x + (-4 - 2x) = 9$$
$$3x + -2x - 4 = 9$$
$$x - 4 = 9$$
$$x - 4 + 4 = 9 + 4$$
$$x = 9 + 4$$
$$x = 13$$

Substitute 13 for x in either equation:

$$2x + y = -4$$
$$2(13) + y = -4$$
$$26 + y = -4$$
$$26 - 26 + y = -4 - 26$$
$$y = -30$$

7. The correct answer is (–0.6, –4.8).

Solve the first equation for y:

$$3x - y = 3$$
$$3x - 3x - y = 3 - 3x$$
$$-y = 3 - 3x$$
$$y = -3 + 3x$$

Substitute $-3 + 3x$ for y in the second equation:

$$4x - 8y = 36$$
$$4x - 8(-3 + 3x) = 36$$
$$4x - 8(3x - 3) = 36$$
$$4x - 24x + 24 = 36$$
$$-20x + 24 = 36$$
$$-20x - 24 + 24 = 36 - 24$$
$$-20x = 12$$
$$\frac{-20x}{-20} = \frac{12}{-20}$$
$$x = \frac{12}{-20}$$
$$x = -0.6$$

Substitute –0.6 for x in either equation:

$$3x - y = 3$$
$$3(-0.6) - y = 3$$
$$-1.8 - y = 3$$
$$-1.8 + 1.8 - y = 3 + 1.8$$
$$-y = 3 + 1.8$$
$$-y = 4.8$$
$$\frac{-y}{-1} = \frac{4.8}{-1}$$
$$y = \frac{4.8}{-1}$$
$$y = -4.8$$

8. The correct answer is (3, 2).

Solve the first equation for x:

$$6y - 2x = 6$$
$$-6y + 6y - 2x = 6 - 6y$$
$$-2x = 6 - 6y$$
$$\frac{-2x}{-2} = \frac{6 - 6y}{-2}$$
$$x = \frac{6 - 6y}{-2}$$
$$x = -3 + 3y$$
$$x = 3y - 3$$

Substitute $3y - 3$ for x in the second equation:

$$3y - 4x = -6$$
$$3y - 4(3y - 3) = -6$$
$$3y - 12y + 12 = -6$$
$$-9y + 12 = -6$$
$$-9y + 12 - 12 = -6 - 12$$
$$-9y = -6 - 12$$
$$-9y = -18$$
$$\frac{-9y}{-9} = \frac{-18}{-9}$$
$$y = \frac{-18}{-9}$$
$$y = 2$$

Substitute 2 for y in either equation:

$$6y - 2x = 6$$
$$6(2) - 2x = 6$$
$$12 - 2x = 6$$
$$12 - 12 - 2x = 6 - 12$$
$$-2x = 6 - 12$$
$$-2x = -6$$
$$\frac{-2x}{-2} = \frac{-6}{-2}$$
$$x = \frac{-6}{-2}$$
$$x = 3$$

9. The correct answer is (2, 8).

Solve the second equation for *y*:

$$y - 3x = 2$$
$$y - 3x + 3x = 2 + 3x$$
$$y = 2 + 3x$$

Substitute 2 + 3*x* for *y* in the first equation:

$$5y + 4x = 48$$
$$5(2 + 3x) + 4x = 48$$
$$10 + 15x + 4x = 48$$
$$10 + 19x = 48$$
$$10 - 10 + 19x = 48 - 10$$
$$19x = 48 - 10$$
$$19x = 38$$
$$\frac{19x}{19} = \frac{38}{19}$$
$$x = \frac{38}{19}$$
$$x = 2$$

Substitute 2 for *x* in either equation:

$$y - 3x = 2$$
$$y - 3(2) = 2$$
$$y - 6 = 2$$
$$y - 6 + 6 = 2 + 6$$
$$y = 2 + 6$$
$$y = 8$$

10. The correct answer is (3, 8).

Solve the first equation for *y*:

$$-3x + y = -1$$
$$-3x + 3x + y = -1 + 3x$$
$$y = -1 + 3x$$

Substitute –1 + 3*x* for *y* in the second equation:

$$-2x + y = 2$$
$$y - 2x = 2$$
$$(-1 + 3x) - 2x = 2$$
$$3x - 1 - 2x = 2$$
$$x - 1 = 2$$
$$x - 1 + 1 = 2 + 1$$
$$x = 2 + 1$$
$$x = 3$$

Substitute 3 for *x* in either equation:

$$-3x + y = -1$$
$$y - 3x = -1$$
$$y - 3(3) = -1$$
$$y - 9 = -1$$
$$y - 9 + 9 = -1 + 9$$
$$y = -1 + 9$$
$$y = 8$$

Solving systems of equations by combining

11. The correct answer is (1, 1).

Subtract the second equation from the first:

$$5x + 6y = 11$$
$$- \ 3x + 6y = 9$$
$$2x + 0y = 2$$

Solve for x:

$$2x + 0y = 2$$
$$2x = 2$$
$$\frac{2x}{2} = \frac{2}{2}$$
$$x = \frac{2}{2}$$
$$x = 1$$

Substitute 1 for x in the first equation:

$$5x + 6y = 11$$
$$5(1) + 6y = 11$$
$$5 + 6y = 11$$
$$5 - 5 + 6y = 11 - 5$$
$$6y = 11 - 5$$
$$6y = 6$$
$$\frac{6y}{6} = \frac{6}{6}$$
$$y = \frac{6}{6}$$
$$y = 1$$

12.

The correct answer is (2, 0).

Add the second equation to the first:

$$7x + 2y = 14$$
$$+ \ 6x - 2y = 12$$
$$13x + 0y = 26$$

Solve for x:

$$13x + 0y = 26$$
$$13x = 26$$
$$\frac{13x}{13} = \frac{26}{13}$$
$$x = \frac{26}{13}$$
$$x = 2$$

Substitute 2 for x in the first equation:

$$7x + 2y = 14$$
$$7(2) + 2y = 14$$
$$14 + 2y = 14$$
$$14 - 14 + 2y = 14 - 14$$
$$2y = 14 - 14$$
$$2y = 0$$
$$\frac{2y}{2} = \frac{0}{2}$$
$$y = \frac{0}{2}$$
$$y = 0$$

Substitute -2 for x in the first equation:

$$4x + 3y = 10$$
$$4(-2) + 3y = 10$$
$$-8 + 3y = 10$$
$$8 + (-8) + 3y = 10 + 8$$
$$8 - 8 + 3y = 10 + 8$$
$$3y = 10 + 8$$
$$3y = 18$$
$$\frac{3y}{3} = \frac{18}{3}$$
$$y = \frac{18}{3}$$
$$y = 6$$

13. The correct answer is $(-2, 6)$.

Subtract the second equation from the first:

$$4x + 3y = 10$$
$$\underline{-\ 2x + 3y = 14}$$
$$2x + 0y = -4$$

Solve for x:

$$2x + 0y = -4$$
$$2x = -4$$
$$\frac{2x}{2} = \frac{-4}{2}$$
$$x = \frac{-4}{2}$$
$$x = -2$$

14. The correct answer is $(1, -1)$.

Begin by multiplying the second equation by 2.

$$2x - 4y = 6$$
$$\underline{\qquad\qquad \times 2}$$
$$4x - 8y = 12$$

Then, subtract the second equation from the first:

$$4x - y = 5$$
$$\underline{-\ 4x - 8y = 12}$$
$$0x + 7y = -7$$

Solve for y:

$$0x + 7y = -7$$
$$7y = -7$$
$$\frac{7y}{7} = \frac{-7}{7}$$
$$y = \frac{-7}{7}$$
$$y = -1$$

Substitute −1 for y in the first equation:

$$4x - y = 5$$
$$4x - (-1) = 5$$
$$4x + 1 = 5$$
$$4x + 1 - 1 = 5 - 1$$
$$4x = 5 - 1$$
$$4x = 4$$
$$\frac{4x}{4} = \frac{4}{4}$$
$$x = \frac{4}{4}$$
$$x = 1$$

15. The correct answer is (10, −2).

Add the second equation to the first:

$$x - y = 12$$
$$+ \ x + y = 8$$
$$\overline{2x + 0y = 20}$$

Solve for x:

$$2x + 0y = 20$$
$$2x = 20$$
$$\frac{2x}{2} = \frac{20}{2}$$
$$x = \frac{20}{2}$$
$$x = 10$$

Substitute 10 for x in the first equation:

$$x - y = 12$$
$$(10) - y = 12$$
$$10 - 10 - y = 12 - 10$$
$$-y = 12 - 10$$
$$-y = 2$$
$$\frac{-y}{-1} = \frac{2}{-1}$$
$$y = \frac{2}{-1}$$
$$y = -2$$

16. The correct answer is $\left(-\frac{1}{2}, \ -\frac{3}{2}\right)$.

Begin by multiplying the first equation by 3.

$$-5x + y = 1$$
$$\underline{ \times 3}$$
$$-15x + 3y = 3$$

Then, add the second equation to the first:

$$-15x + 3y = 3$$
$$+ \ 9x - 3y = 0$$
$$\overline{-6x + 0y = 3}$$

Solve for x:

$$-6x + 0y = 3$$
$$-6x = 3$$
$$\frac{-6x}{-6} = \frac{3}{-6}$$
$$x = \frac{3}{-6}$$
$$x = -\frac{1}{2}$$

Substitute $-\frac{1}{2}$ for x in the first equation:

$$-5x + y = 1$$
$$-5\left(-\frac{1}{2}\right) + y = 1$$
$$\frac{5}{2} + y = 1$$
$$\frac{5}{2} - \frac{5}{2} + y = 1 - \frac{5}{2}$$
$$y = 1 - \frac{5}{2}$$
$$y = \frac{2}{2} - \frac{5}{2}$$
$$y = -\frac{3}{2}$$

17. The correct answer is $(5, -1)$.

Begin by multiplying the first equation by 2.

$$3x - 2y = 17$$
$$\underline{\qquad\quad \times\ 2}$$
$$6x - 4y = 34$$

Also, multiply the second equation by 3.

$$-2x - 5y = -5$$
$$\underline{\qquad\quad \times\ 3}$$
$$-6x - 15y = -15$$

Then, add the second equation to the first:

$$6x - 4y = 34$$
$$\underline{+ - 6x - 15y = -15}$$
$$0x - 19y = 19$$

Solve for y:

$$0x - 19y = 19$$
$$-19y = 19$$
$$\frac{-19y}{-19} = \frac{19}{-19}$$
$$y = \frac{19}{-19}$$
$$y = -1$$

Substitute -1 for y in the first equation:

$$3x - 2y = 17$$
$$3x - 2(-1) = 17$$
$$3x + 2 = 17$$
$$3x + 2 - 2 = 17 - 2$$
$$3x = 17 - 2$$
$$3x = 15$$
$$\frac{3x}{3} = \frac{15}{3}$$
$$x = \frac{15}{3}$$
$$x = 5$$

18. The correct answer is $(-1, 0)$.

Begin by multiplying the first equation by 2.

$$-2x + y = 2$$
$$\underline{\qquad\quad \times\ 2}$$
$$-4x + 2y = 4$$

Then, subtract the second equation from the first:

$$-4x + 2y = 4$$
$$\underline{- -4x + 4y = 4}$$
$$0x - 2y = 0$$

Solve for y:

$$0x - 2y = 0$$
$$-2y = 0$$
$$\frac{-2y}{-2} = \frac{0}{-2}$$
$$y = \frac{0}{-2}$$
$$y = 0$$

Substitute 0 for y in the first equation:

$$-2x + y = 2$$
$$-2x + (0) = 2$$
$$-2x = 2$$
$$\frac{-2x}{-2} = \frac{2}{-2}$$
$$x = \frac{2}{-2}$$
$$x = -1$$

Substitute $-\frac{1}{5}$ for x in the first equation:

$$3x + 3y = 0$$
$$3\left(-\frac{1}{5}\right) + 3y = 0$$
$$-\frac{3}{5} + 3y = 0$$
$$-\frac{3}{5} + \frac{3}{5} + 3y = 0 + \frac{3}{5}$$
$$3y = 0 + \frac{3}{5}$$
$$3y = \frac{3}{5}$$
$$\frac{3y}{3} = \frac{3}{5} \times \frac{1}{3}$$
$$y = \frac{3}{5} \times \frac{1}{3}$$
$$y = \frac{3}{15}$$
$$y = \frac{1}{5}$$

19. The correct answer is $\left(-\frac{1}{5}, \frac{1}{5}\right)$.

Begin by multiplying the second equation by 3.

$$4x - y = -1$$
$$\underline{\times\ 3}$$
$$12x - 3y = -3$$

Then, add the second equation to the first:

$$3x + 3y = 0$$
$$\underline{+\ 12x - 3y = -3}$$
$$15x + 0y = -3$$

Solve for x:

$$15x + 0y = -3$$
$$15x = -3$$
$$\frac{15x}{15} = \frac{-3}{15}$$
$$x = \frac{-3}{15}$$
$$x = -\frac{1}{5}$$

20.

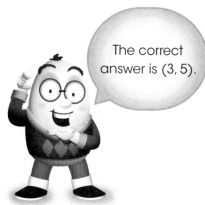

The correct answer is (3, 5).

Chapter Review Solutions

Add the second equation to the first:

$$3x + y = 14$$
$$\underline{+\ x - y = -2}$$
$$4x + 0y = 12$$

Solve for x:

$$4x + 0y = 12$$
$$4x = 12$$
$$\frac{4x}{4} = \frac{12}{4}$$
$$x = \frac{12}{4}$$
$$x = 3$$

Substitute 3 for x in the first equation:

$$3x + y = 14$$
$$3(3) + y = 14$$
$$9 + y = 14$$
$$9 - 9 + y = 14 - 9$$
$$y = 14 - 9$$
$$y = 5$$

1. The correct answer is (−3, 3).

Solve the second equation for x:

$$x + 2y = 3$$
$$x + 2y - 2y = 3 - 2y$$
$$x = 3 - 2y$$

Substitute $3 - 2y$ for x in the first equation:

$$5x + y = -12$$
$$5(3 - 2y) + y = -12$$
$$15 - 10y + y = -12$$
$$15 - 9y = -12$$
$$15 - 15 - 9y = -12 - 15$$
$$-9y = -12 - 15$$
$$-9y = -27$$
$$\frac{-9y}{-9} = \frac{-27}{-9}$$
$$y = \frac{-27}{-9}$$
$$y = 3$$

Substitute 3 for y in either equation:

$$x + 2y = 3$$
$$x + 2(3) = 3$$
$$x + 6 = 3$$
$$x + 6 - 6 = 3 - 6$$
$$x = 3 - 6$$
$$x = -3$$

2.

The correct answer is $(2, 6)$.

Solve the second equation for y:

$$3x - y = 0$$
$$3x - 3x - y = 0 - 3x$$
$$-y = 0 - 3x$$
$$-y = -3x$$
$$\frac{-y}{1} = \frac{-3x}{-1}$$
$$y = 3x$$

Substitute $3x$ for y in the first equation:

$$2x - 4y = -20$$
$$2x - 4(3x) = -20$$
$$2x - 12x = -20$$
$$-10x = -20$$
$$\frac{-10x}{-10} = \frac{-20}{-10}$$
$$x = \frac{-20}{-10}$$
$$x = 2$$

Substitute 2 for x in either equation:

$$3x - y = 0$$
$$3(2) - y = 0$$
$$6 - y = 0$$
$$6 - 6 - y = 0 - 6$$
$$-y = -6$$
$$\frac{-y}{-1} = \frac{-6}{-1}$$
$$y = \frac{-6}{-1}$$
$$y = 6$$

3. The correct answer is $(-1, 1)$.

Solve the first equation for x:

$$x + 3y = 2$$
$$x + 3y - 3y = 2 - 3y$$
$$x = 2 - 3y$$

Substitute $2 - 3y$ for x in the second equation:

$$-2x + 4y = 6$$
$$-2(2 - 3y) + 4y = 6$$
$$-4 + 6y + 4y = 6$$
$$-4 + 10y = 6$$
$$-4 + 4 + 10y = 6 + 4$$
$$10y = 6 + 4$$
$$10y = 10$$
$$\frac{10y}{10} = \frac{10}{10}$$
$$y = \frac{10}{10}$$
$$y = 1$$

Substitute 1 for y in either equation:

$$x + 3y = 2$$
$$x + 3(1) = 2$$
$$x + 3 = 2$$
$$x + 3 - 3 = 2 - 3$$
$$x = 2 - 3$$
$$x = -1$$

4. The correct answer is (–4, –7).

   Solve the second equation for y:

$$y - 2x = 1$$
$$y - 2x + 2x = 1 + 2x$$
$$y = 1 + 2x$$

   Substitute $1 + 2x$ for y in the first equation:

$$2y - 3x = -2$$
$$2(1 + 2x) - 3x = -2$$
$$2 + 4x - 3x = -2$$
$$4x - 3x + 2 = -2$$
$$x + 2 = -2$$
$$x + 2 - 2 = -2 - 2$$
$$x = -2 - 2$$
$$x = -4$$

   Substitute –4 for x in either equation:

$$y - 2x = 1$$
$$y - 2(-4) = 1$$
$$y + 8 = 1$$
$$y + 8 - 8 = 1 - 8$$
$$y = 1 - 8$$
$$y = -7$$

5. The correct answer is (–1, 0).

   Solve the first equation for y:

$$y - x = 1$$
$$y - x + x = 1 + x$$
$$y = 1 + x$$

   Substitute $1 + x$ for y in the second equation:

$$4y - 8x = 8$$
$$4(1 + x) - 8x = 8$$
$$4 + 4x - 8x = 8$$
$$4 - 4x = 8$$
$$-4x + 4 = 8$$
$$-4x + 4 - 4 = 8 - 4$$
$$-4x = 4$$
$$\frac{-4x}{-4} = \frac{4}{-4}$$
$$x = -1$$

   Substitute –1 for x in either equation:

$$y - x = 1$$
$$y - (-1) = 1$$
$$y + 1 = 1$$
$$y + 1 - 1 = 1 - 1$$
$$y = 0$$

6. The correct answer is (2, 9).

   Solve the first equation for y:

$$5x - y = 1$$
$$5x - 5x - y = 1 - 5x$$
$$-y = 1 - 5x$$
$$\frac{-y}{-1} = \frac{1 - 5x}{-1}$$
$$y = \frac{1 - 5x}{-1}$$
$$y = -1 + 5x$$
$$y = 5x - 1$$

Substitute $5x - 1$ for y in the second equation:

$$3x - 2y = -12$$
$$3x - 2(5x - 1) = -12$$
$$3x - 10x + 2 = -12$$
$$-7x + 2 = -12$$
$$-7x + 2 - 2 = -12 - 2$$
$$-7x = -14$$
$$\frac{-7x}{-7} = \frac{-14}{-7}$$
$$x = \frac{-14}{-7}$$
$$x = 2$$

Substitute 2 for x in either equation:

$$5x - y = 1$$
$$5(2) - y = 1$$
$$10 - y = 1$$
$$10 - 10 - y = 1 - 10$$
$$-y = 1 - 10$$
$$-y = -9$$
$$\frac{-y}{-1} = \frac{-9}{-1}$$
$$y = 9$$

7. The correct answer is (–7, –2).

Solve the first equation for x:

$$x - 3y = -1$$
$$x - 3y + 3y = -1 + 3y$$
$$x = -1 + 3y$$

Substitute $-1 + 3y$ for x in the second equation:

$$x - y = -5$$
$$(-1 + 3y) - y = -5$$
$$-1 + 2y = -5$$
$$-1 + 1 + 2y = -5 + 1$$
$$2y = -5 + 1$$
$$2y = -4$$
$$\frac{2y}{2} = \frac{-4}{2}$$
$$y = \frac{-4}{2}$$
$$y = -2$$

Substitute –2 for y in either equation:

$$x - 3y = -1$$
$$x - 3(-2) = -1$$
$$x + 6 = -1$$
$$x + 6 - 6 = -1 - 6$$
$$x = -7$$

8. The correct answer is (14, 4).

Solve the first equation for x:

$$x - 3y = 2$$
$$x - 3y + 3y = 2 + 3y$$
$$x = 2 + 3y$$

Substitute 2 + 3*y* for *x* in the second equation:

$$2x + 6y = 52$$
$$2(2 + 3y) + 6y = 52$$
$$4 + 6y + 6y = 52$$
$$4 + 12y = 52$$
$$4 - 4 + 12y = 52 - 4$$
$$12y = 48$$
$$\frac{12y}{12} = \frac{48}{12}$$
$$y = \frac{48}{12}$$
$$y = 4$$

Substitute 4 for *y* in either equation:

$$x - 3y = 2$$
$$x - 3(4) = 2$$
$$x - 12 = 2$$
$$x - 12 + 12 = 2 + 12$$
$$x = 2 + 12$$
$$x = 14$$

9. The correct answer is (4, 6).

Solve the first equation for *y:*

$$y - x = 2$$
$$y - x + x = 2 + x$$
$$y = 2 + x$$

Substitute 2 + *x* for *y* in the second equation:

$$2y - 3x = 0$$
$$2(2 + x) - 3x = 0$$
$$4 + 2x - 3x = 0$$
$$4 - x = 0$$
$$4 - 4 - x = 0 - 4$$
$$-x = 0 - 4$$
$$-x = -4$$
$$\frac{-x}{-1} = \frac{-4}{-1}$$
$$x = 4$$

Substitute 4 for *x* in either equation:

$$y - x = 2$$
$$y - (4) = 2$$
$$y - 4 + 4 = 2 + 4$$
$$y = 2 + 4$$
$$y = 6$$

10. The correct answer is (−6, −4).

Solve the second equation for *x:*

$$4x - 12y = 24$$
$$4x - 12y + 12y = 24 + 12y$$
$$4x = 24 + 12y$$
$$\frac{4x}{4} = \frac{24 + 12y}{4}$$
$$x = \frac{24}{4} + \frac{12y}{4}$$
$$x = 6 + 3y$$

Substitute $6 + 3y$ for x in the first equation:

$$2x - 3y = 0$$
$$2(6 + 3y) - 3y = 0$$
$$12 + 6y - 3y = 0$$
$$12 + 3y = 0$$
$$12 - 12 + 3y = 0 - 12$$
$$3y = 0 - 12$$
$$3y = -12$$
$$\frac{3y}{3} = \frac{-12}{3}$$
$$y = \frac{-12}{3}$$
$$y = -4$$

Substitute -4 for y in either equation:

$$4x - 12y = 24$$
$$4x - 12(-4) = 24$$
$$4x + 48 = 24$$
$$4x + 48 - 48 = 24 - 48$$
$$4x = 24 - 48$$
$$4x = -24$$
$$\frac{4x}{4} = \frac{-24}{4}$$
$$x = \frac{-24}{4}$$
$$x = -6$$

11. The correct answer is $(-12, -7)$.

   Solve the first equation for y:

$$-x + y = 5$$
$$x - x + y = 5 + x$$
$$y = 5 + x$$

Substitute $5 + x$ for y in the second equation:

$$x - 2y = 2$$
$$x - 2(5 + x) = 2$$
$$x - 10 - 2x = 2$$
$$-x - 10 = 2$$
$$-x - 10 + 10 = 2 + 10$$
$$-x = 2 + 10$$
$$-x = 12$$
$$\frac{-x}{-1} = \frac{12}{-1}$$
$$x = -12$$

Substitute -12 for x in either equation:

$$-x + y = 5$$
$$-(-12) + y = 5$$
$$12 + y = 5$$
$$12 - 12 + y = 5 - 12$$
$$y = 5 - 12$$
$$y = -7$$

12. The correct answer is $(0, 2)$.

   Solve the first equation for y:

$$y - x = 2$$
$$y - x + x = 2 + x$$
$$y = 2 + x$$

Substitute $2 + x$ for y in the second equation:

$$3x + 5y = 10$$
$$3x + 5(2 + x) = 10$$
$$3x + 10 + 5x = 10$$
$$8x + 10 = 10$$
$$8x + 10 - 10 = 10 - 10$$
$$8x = 0$$
$$\frac{8x}{8} = \frac{0}{8}$$
$$x = \frac{0}{8}$$
$$x = 0$$

Substitute 0 for x in either equation:

$$y - x = 2$$
$$y - (0) = 2$$
$$y = 2$$

13. The correct answer is $(-10, -16)$.

Solve the first equation for y:

$$3y - 6x = 12$$
$$3y - 6x + 6x = 12 + 6x$$
$$3y = 12 + 6x$$
$$\frac{3y}{3} = \frac{12 + 6x}{3}$$
$$y = \frac{12 + 6x}{3}$$
$$y = \frac{12}{3} + \frac{6x}{3}$$
$$y = 4 + 2x$$

Substitute $4 + 2x$ for y in the second equation:

$$2y - 3x = -2$$
$$2(4 + 2x) - 3x = -2$$
$$8 + 4x - 3x = -2$$
$$8 + x = -2$$
$$8 - 8 + x = -2 - 8$$
$$x = -10$$

Substitute -10 for x in either equation:

$$3y - 6x = 12$$
$$3y - 6(-10) = 12$$
$$3y + 60 = 12$$
$$3y + 60 - 60 = 12 - 60$$
$$3y = 12 - 60$$
$$3y = -48$$
$$\frac{3y}{3} = \frac{-48}{3}$$
$$y = \frac{-48}{3}$$
$$y = -16$$

14. The correct answer is $(-4, 5)$.

Subtract the second equation from the first:

$$x + y = 1$$
$$\underline{- x + 5y = 21}$$
$$0x - 4y = -20$$

Solve for y:

$$0x - 4y = -20$$
$$-4y = -20$$
$$\frac{-4y}{-4} = \frac{-20}{-4}$$
$$y = \frac{-20}{-4}$$
$$y = 5$$

Substitute 5 for y in the first equation:

$$x + y = 1$$
$$x + (5) = 1$$
$$x + 5 - 5 = 1 - 5$$
$$x = 1 - 5$$
$$x = -4$$

15. The correct answer is (5, –2).

Begin by multiplying the second equation by 3:

$$x + 6y = -7$$
$$\underline{ \times\ 3}$$
$$3x + 18y = -21$$

Then, subtract the second equation from the first:

$$3x + y = 13$$
$$\underline{-\ 3x + 18y = -21}$$
$$0x - 17y = 34$$

Solve for y:

$$0x - 17y = 34$$
$$-17y = 34$$
$$\frac{-17y}{-17} = \frac{34}{-17}$$
$$y = \frac{34}{-17}$$
$$y = -2$$

Substitute –2 for y in the second equation:

$$x + 6y = -7$$
$$x + 6(-2) = -7$$
$$x - 12 = -7$$
$$x - 12 + 12 = -7 + 12$$
$$x = -7 + 12$$
$$x = 5$$

16. The correct answer is (1, 2).

Begin by multiplying the first equation by 2:

$$5x - y = 3$$
$$\underline{ \times\ 2}$$
$$10x - 2y = 6$$

Then, add the second equation to the first:

$$10x - 2y = 6$$
$$\underline{+\ \ -10x + 5y = 0}$$
$$0x + 3y = 6$$

Solve for y:

$$0x + 3y = 6$$
$$3y = 6$$
$$\frac{3y}{3} = \frac{6}{3}$$
$$y = \frac{6}{3}$$
$$y = 2$$

Substitute 2 for y in the first equation:

$$5x - y = 3$$
$$5x - (2) = 3$$
$$5x - 2 + 2 = 3 + 2$$
$$5x = 3 + 2$$
$$5x = 5$$
$$\frac{5x}{5} = \frac{5}{5}$$
$$x = \frac{5}{5}$$
$$x = 1$$

17. The correct answer is (6, 1).

Add the second equation to the first:

$$x + 2y = 8$$
$$\underline{+\ x - 2y = 4}$$
$$2x + 0y = 12$$

Solve for x:

$$2x + 0y = 12$$
$$2x = 12$$
$$\frac{2x}{2} = \frac{12}{2}$$
$$x = \frac{12}{2}$$
$$x = 6$$

Substitute 6 for x in the first equation:

$$x + 2y = 8$$
$$(6) + 2y = 8$$
$$6 - 6 + 2y = 8 - 6$$
$$2y = 8 - 6$$
$$2y = 2$$
$$\frac{2y}{2} = \frac{2}{2}$$
$$y = \frac{2}{2}$$
$$y = 1$$

18.

The correct answer is (0, 2).

Subtract the second equation from the first:

$$4x + 7y = 14$$
$$\underline{-\ 4x - 5y = -10}$$
$$0x + 12y = 24$$

Solve for y:

$$0x + 12y = 24$$
$$12y = 24$$
$$\frac{12y}{12} = \frac{24}{12}$$
$$y = \frac{24}{12}$$
$$y = 2$$

Substitute 2 for y in the second equation:

$$4x - 5y = -10$$
$$4x - 5(2) = -10$$
$$4x - 10 = -10$$
$$4x - 10 + 10 = -10 + 10$$
$$4x = -10 + 10$$
$$4x = 0$$
$$\frac{4x}{4} = \frac{0}{4}$$
$$x = \frac{0}{4}$$
$$x = 0$$

19. The correct answer is (5, 4).

Begin by multiplying the first equation by 2:

$$2x - 3y = -2$$
$$\underline{\times\ 2}$$
$$4x - 6y = -4$$

Then, subtract the second equation from the first:

$$4x - 6y = -4$$
$$\underline{-4x + y = 24}$$
$$0x - 7y = -28$$

Solve for y:

$$0x - 7y = -28$$
$$-7y = -28$$
$$\frac{-7y}{-7} = \frac{-28}{-7}$$
$$y = \frac{-28}{-7}$$
$$y = 4$$

Substitute 4 for y in the first equation:

$$2x - 3y = -2$$
$$2x - 3(4) = -2$$
$$2x - 12 = -2$$
$$2x - 12 + 12 = -2 + 12$$
$$2x = -2 + 12$$
$$2x = 10$$
$$\frac{2x}{2} = \frac{10}{2}$$
$$x = \frac{10}{2}$$
$$x = 5$$

20. The correct answer is (3, 9).

Subtract the second equation from the first:

$$9x + y = 36$$
$$\underline{-x + y = 12}$$
$$8x - 0y = 24$$

Solve for x:

$$8x - 0y = 24$$
$$8x = 24$$
$$\frac{8x}{8} = \frac{24}{8}$$
$$x = \frac{24}{8}$$
$$x = 3$$

Substitute 3 for x in the second equation:

$$x + y = 12$$
$$(3) + y = 12$$
$$3 - 3 + y = 12 - 3$$
$$y = 12 - 3$$
$$y = 9$$

21. The correct answer is (−7, 31).

Begin by multiplying the second equation by 2:

$$x + y = 24$$
$$\underline{\times\ 2}$$
$$2x + 2y = 48$$

Then, subtract the second equation from the first:

$$7x + 2y = 13$$
$$\underline{-2x + 2y = 48}$$
$$5x + 0y = -35$$

Solve for x:

$$5x + 0y = -35$$
$$5x = -35$$
$$\frac{5x}{5} = \frac{-35}{5}$$
$$x = \frac{-35}{5}$$
$$x = -7$$

Substitute –7 for x in the second equation:

$$x + y = 24$$
$$(-7) + y = 24$$
$$-7 + y = 24$$
$$-7 + 7 + y = 24 + 7$$
$$y = 24 + 7$$
$$y = 31$$

22. The correct answer is (5, –1).

Begin by multiplying the first equation by 2:

$$2x + y = 9$$
$$\underline{\times\ 2}$$
$$4x + 2y = 18$$

Then, add the second equation to the first:

$$4x + 2y = 18$$
$$\underline{+\ 6x - 2y = 32}$$
$$10x - 0y = 50$$

Solve for x:

$$10x - 0y = 50$$
$$10x = 50$$
$$\frac{10x}{10} = \frac{50}{10}$$
$$x = \frac{50}{10}$$
$$x = 5$$

Substitute 5 for x in the first equation:

$$2x + y = 9$$
$$2(5) + y = 9$$
$$10 + y = 9$$
$$10 - 10 + y = 9 - 10$$
$$y = 9 - 10$$
$$y = -1$$

23. The correct answer is (6, 0).

Begin by multiplying the first equation by 4:

$$2x - y = 12$$
$$\underline{\times\ 4}$$
$$8x - 4y = 48$$

Then, add the second equation to the first:

$$8x - 4y = 48$$
$$\underline{+\ 3x + 4y = 18}$$
$$11x + 0y = 66$$

Solve for x:

$$11x + 0y = 66$$
$$11x = 66$$
$$\frac{11x}{11} = \frac{66}{11}$$
$$x = \frac{66}{11}$$
$$x = 6$$

Substitute 6 for x in the first equation:

$$2x - y = 12$$
$$2(6) - y = 12$$
$$12 - y = 12$$
$$-12 + 12 - y = 12 - 12$$
$$-y = 0$$
$$\frac{-y}{-1} = \frac{0}{-1}$$
$$y = 0$$

24. The correct answer is (10, 5).

Begin by multiplying the first equation by 3:

$$4x - 3y = 25$$
$$\underline{\times\ 3}$$
$$12x - 9y = 75$$

Then, multiply the second equation by 4:

$$-3x + 8y = 10$$
$$\underline{\times\ 4}$$
$$-12x + 32y = 40$$

Then, add the second equation to the first:

$$12x - 9y = 75$$
$$\underline{+ -\ 12x + 32y = 40}$$
$$0x + 23y = 115$$

Solve for y:

$$0x + 23y = 115$$
$$23y = 115$$
$$\frac{23y}{23} = \frac{115}{23}$$
$$y = \frac{115}{23}$$
$$y = 5$$

Substitute 5 for y in the first equation:

$$4x - 3y = 25$$
$$4x - 3(5) = 25$$
$$4x - 15 = 25$$
$$4x - 15 + 15 = 25 + 15$$
$$4x = 25 + 15$$
$$4x = 40$$
$$\frac{4x}{4} = \frac{40}{4}$$
$$x = \frac{40}{4}$$
$$x = 10$$

25. The correct answer is (2, –2).

Subtract the second equation from the first:

$$x - 2y = 6$$
$$\underline{-\ 4x - 2y = 12}$$
$$-3x + 0y = -6$$

Solve for x:

$$-3x + 0y = -6$$
$$-3x = -6$$
$$\frac{-3x}{-3} = \frac{-6}{-3}$$
$$x = \frac{-6}{-3}$$
$$x = 2$$

Substitute 2 for x in the second equation:

$$4x - 2y = 12$$
$$4(2) - 2y = 12$$
$$8 - 2y = 12$$
$$8 - 8 - 2y = 12 - 8$$
$$-2y = 12 - 8$$
$$-2y = 4$$
$$\frac{-2y}{-2} = \frac{4}{-2}$$
$$y = \frac{4}{-2}$$
$$y = -2$$

Chapter 11

Word Problems

**In this chapter, we'll review
the following concepts:**

What is a word problem?
Types of word problems
Setting up word problems
Solving word problems

What is a word problem?

An algebra word problem is a scenario that is described in English. It requests missing information that can be determined using math. To solve a word problem, you must translate the description into a math statement and calculate the answer.

Example

Tori is 5 years older than her sister Laney. If Laney is 13, how old is Tori?

In the word problem above, we're given some information about Tori and Laney's ages. We must use the information to create a math statement that lets us determine Tori's exact age.

Types of word problems

Word problems can require you to use your knowledge of multiple areas of math.

Here are some of the most common types.

Integers

Some word problems simply require you to find missing numbers. The example above is one such problem. You are given some information about the ages of Laney and Tori and asked to find Tori's age.

egghead's Guide to Algebra

Ratios and proportions

Word problems may present information in terms of a ratio. You must then set up a proportion to find a missing value.

Eric writes thank-you notes for the gifts he received at graduation. He has 30 thank-you notes to write. He writes 10 thank-you notes in the first 2 hours. At that rate, how long will it take Eric to write all 30 notes?

This problem starts with the rate at which Eric writes thank-you notes. This rate can be expressed as a **ratio**: 10 notes for every 2 hours, or 10:2. The ratio can also be written as a fraction: $\frac{10}{2}$.

To solve this problem, you would need to set up a proportion. A **proportion** is an equation containing two ratios.

$$\frac{10 \text{ notes}}{2 \text{ hours}} = \frac{30 \text{ notes}}{\text{total hours}}$$

You are looking for the total number of hours it takes Eric to write all 30 notes. We'll see how to find this value below.

Time and distance

Time and distance are also commonly addressed in word problems. You might see problems involving travel, for instance. To solve these problems, we use the following formula:

Distance = rate × time

The rate of travel is usually given in miles per hour or kilometers per hour. The formula can also therefore be written as distance = speed × time:

$$D = s \times t$$

Measurements and conversions

Some word problems address measurements in miles, kilometers, yards, meters, feet, inches, or centimeters. You may be required to convert between units, such as changing inches to feet.

Two common measurement problems involve finding the perimeter and area of shapes. The **perimeter** of a figure is the distance around its outer edge. To find the perimeter of a rectangle, for instance, we add up length + length + width + width:

$$\text{Perimeter} = (2 \times \text{length}) + (2 \times \text{width})$$

The **area** of a figure is the amount of space that it covers. To find the area of a rectangle, we multiply length × width:

$$\text{Area} = \text{length} \times \text{width}$$

These formulas are important for solving measurement problems.

Cost and percent

Another common type of algebra word problem involves cost and percent.

Andre buys a jacket at 20% off the regular price. The regular price of the jacket is $100. What is the sale price?

In this example, we must use our knowledge of percents to determine the price Andre pays. When working with percents, remember that percents can be expressed as a number divided by 100:

$$20\% = \frac{20}{100}$$

The fraction could also be expressed as a decimal number:

$$20\% = 0.20$$

When converting from percents to decimals, to avoid having to divide, you can simply move the decimal point two places to the left of the original number:

20.00% = 0.20

Let's set up some different word problems next.

Setting up word problems

To solve a word problem, you must first translate the description into a math statement. This process is called **setting up** the problem.

When we set up a word problem, we write it in mathematical terms. This may involve creating an equation or an inequality. We may need a linear equation, a quadratic, or even a cubic equation. We may create single equations or systems of equations, depending on the scenario described.

All of the information you need to solve the problem is given in the description. Your job is to create the correct math statements and calculate to find the answer.

Examples

Let's start with the integer problem we saw above.

Tori is 5 years older than her sister Laney. If Laney is 13, how old is Tori?

To set up this problem, we can create an equation. Use variables to represent the missing numbers.

In this case, let t represent Tori's age. Let l represent Laney's age. The description tells us that Tori is 5 years older than Laney. Write this as an equation:

$$t = l + 5$$

We're also told that Laney is 13 years old, so the value of l is 13. Substitute 13 for l in the equation:

$$t = (13) + 5$$

This equation is set up and ready to solve.

As another example, let's try a problem involving time and distance.

A train departs from Station A at 7:00 a.m. It travels to Station B at a constant rate of 60 miles per hour. The train arrives at Station B at 9:00 a.m. How far, in miles, is Station B from Station A?

This word problem asks you to determine the distance traveled by the train. To set up this problem, we'll let the variable d represent the distance traveled. We'll also create an equation based on the distance formula:

$$d = s \times t$$
$$d = (60) \times (2)$$

We know that the train traveled at 60 miles per hour, so we substitute 60 for the train's speed. The train traveled for 2 hours total, so we substitute 2 for the travel time. This word problem is now set up and ready to solve.

Let's also set up a
percent problem.

Andre buys a jacket at 20% off the regular price. The regular price of the jacket is $100. What is the sale price?

To set up this percent problem we saw earlier, we must create an equation. The sale price is equal to the regular price minus the discount. You can write that out first, if it's helpful:

sale price = regular price – amount of discount

Let *s* represent the sale price of the jacket. Substitute *s* for the sale price and $100 for the regular price into the equation:

$s = \$100 -$ amount of discount

To finish setting up the problem, we must represent the amount of the discount. The discount is 20 percent of the original price. That can be written as $\frac{20}{100} \times \$100$, or just $0.20 \times \$100$:

$s = \$100 - (0.20 \times \$100)$

We can go one step further and write:

$s = 100 - 0.20(100)$

The equation is now set up and ready to solve.

Practice Questions—Setting up word problems

Directions: Set up the word problems below. You will find the Practice Question Solutions on page 252.

1. Marek has 17 baseball cards. He gives away 2 of his cards and buys 8 more. How many cards does he have?

2. An ice cream truck sells ice cream bars in the Tannery neighborhood at a rate of 12 per hour. After driving through the Tannery neighborhood for 4 hours, how many ice cream bars will the truck have sold?

3. It takes Marta exactly 15 minutes to walk to school. If Marta walks at a constant speed of 3 miles per hour, how far is the school from her house?

4. James and Ryan play 24 video games in a row. If James wins 37.5 percent of the games they play, how many of the games does Ryan win?

5. Mrs. Lopez buys 3 gallons of milk for $1.99 per gallon at the store. She also buys 2 loaves of bread. If the total amount of her order is $10.11 before tax, what is the cost of each loaf of bread?

6. Owen's bedroom is in the shape of a rectangle. The length measures 9 feet, and the width measures 13.5 feet. What is the area of Owen's room?

7. Kelly spent 3.5 hours at the library studying. She then walked 15 minutes home and took her dog to the park. The total amount of time she spent at the library was twice the amount of time she spent walking home and taking her dog to the park. How long did Kelly spend taking her dog to the park?

egghead's Guide to Algebra

8. Connor weighs three packages to send to his cousins as holiday gifts. The first package weighs 23 ounces. The second package weighs 34 ounces, and the third weighs 17 ounces. What is the weight, in pounds, of all three packages?

10. A jet travels from City A to City B at 200 miles per hour. It then travels from City B back to City A at 400 miles per hour. If the jet makes the entire trip in 7.5 hours, what is the distance from City A to City B?

9. A squirrel stores acorns for the winter at a rate of 1 acorn every 5 minutes. If the squirrel works continuously at the same rate, how many acorns will the squirrel store in one hour and 15 minutes?

Solving word problems

Now that we've reviewed how to set up word problems, let's work on solving them.

Setting them up
is sometimes the
hardest part!

Once we have set up an equation, the next step is to perform the math to solve the problem. For some problems, this may be just a matter of adding, subtracting, multiplying, or dividing.

A train departs from Station A at 7:00 a.m. It travels to Station B at a constant rate of 60 miles per hour. The train arrives at Station B at 9:00 a.m. How far, in miles, is Station B from Station A?

To set up this problem above, we created an equation based on the distance formula:

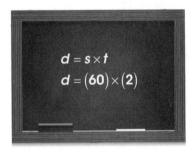

$$d = s \times t$$
$$d = (60) \times (2)$$

To solve the problem, we just need to multiply:

$$d = s \times t$$
$$d = (60) \times (2)$$
$$d = 120$$

Station B is 120 miles from Station A.

Other problems may require more complex math. Remember to follow the order of operations!

Andre buys a jacket at 20% off the regular price. The regular price of the jacket is $100. What is the sale price?

To set up this problem, we created an equation for the amount of the sale price:

$$s = 100 - 0.20(100)$$

To solve for s, perform multiplication first and then subtraction:

$$s = 100 - 0.20(100)$$
$$s = 100 - 20$$
$$s = 80$$

The sale price of the jacket is $80.00.

Other problems may involve solving quadratic equations, solving systems of equations, or finding square roots. Here is an example of a problem using proportions:

Eric writes thank-you notes for the gifts he received at graduation. He has 30 thank-you notes to write. He writes 10 thank-you notes in the first 2 hours. At that rate, how long will it take Eric to write all 30 notes?

To set up this problem, we created a proportion. The proportion expresses a relationship between the two ratios given:

$$\frac{10 \text{ notes}}{2 \text{ hours}} = \frac{30 \text{ notes}}{\text{total hours}}$$

To solve the problem, we now would cross multiply. Let t represent the total hours it would take Eric to write 30 notes:

$$\frac{10 \text{ notes}}{2 \text{ hours}} = \frac{30 \text{ notes}}{\text{total hours}}$$
$$\frac{10}{2} = \frac{30}{t}$$
$$10 \times t = 2 \times 30$$
$$10t = 2 \times 30$$
$$10t = 60$$

Divide both sides by 10 to solve for *t*:

$$10t = 60$$

$$\frac{10t}{10} = \frac{60}{10}$$

$$t = \frac{60}{10}$$

$$t = 6$$

It would take Eric 6 hours to write 30 thank-you notes.

Practice Questions—Solving word problems

Directions: Solve the word problems below. You will find the Practice Question Solutions on page 255.

11. It took Jesse 72 minutes longer to finish his homework than it took Amber to finish hers. The amount of time that Jessie spent was 64 minutes longer than twice the length of time Amber spent. How long did each of them take to finish their homework?

12. Sara buys a paint set on sale for $21.75. The sale price is 25% off of the regular price. What is the regular price of the paint set?

13. The average speed for a competitive cyclist is 30 kilometers per hour. If Jake trains by riding his bike for 2 hours per day at the average competitive speed, how long should his training route be?

14. Train 1 leaves the city at 8:00 a.m. Train 2 leaves the city at noon on a parallel track. The second train's speed is 144 mph faster than the speed of the first train. At what speeds are each of the trains moving if the second train passes the first train at 5:00 p.m.?

15. Geoffrey received a Christmas bonus check of $1,140. He divided it between a standard savings account and a certificate of deposit (CD). The savings account pays interest of 1.4% per year, while the CD earns interest at a rate of 3% per year. At the end of the first year, the two accounts had accumulated the same amount of interest. How much was initially invested in each account?

16. Anna sells candles and earns $4.00 on each candle sold. In addition, she earns a bonus of $62 if she sells more than 50 candles in the week. If Anna needs to earn a minimum of $327 this week in order to make her car payment, how many candles does she need to sell?

17. To qualify for his brown belt in Tae Kwon Do, Drake must master 76 martial arts moves. He has already mastered 21 moves. If he masters 5 additional moves per month, how many months will it take Drake to earn the belt?

18. A woman places bets on two horse races. She bets on a horse named Sweet Tea in the first race, with odds to win of 2:1. She bets on a horse named Oil Slick in the second race, with odds to win of 4:1. If Sweet Tea wins, the payout will be twice the amount of the bet placed. If Oil Slick wins, the payout will be four times the amount of the bet placed. The woman starts with $42 in betting money, and she wins a total of $122. How much does she bet on each horse, if both horses win their races?

19. Brendan spent Saturday playing a video game. It takes a total of 720 points to advance to each new level of the game. Brendan started the day at Level 1 with 347 points. He ended the day with 2,468 points. What level had Brendan achieved when he stopped playing on Saturday?

20. Ashley needs to save $890 to buy a new computer. She has already saved $190 toward the purchase. If she earns $8 per hour, how many hours and minutes must she work to earn exactly enough to pay for the new computer?

Chapter Review

Directions: Set up or solve the word problems below. Solutions can be found on page 259.

1. Janet is 12 years older than Robin. Robin is 4 years older than Stuart. If Stuart is 17, write an equation that will enable you to find Janet's age.

2. Mr. Spector buys a model kit for 30% off the regular price. The regular price of the kit is $17.00. Write an equation that shows how much Mr. Spector paid for the kit.

3. A furniture manufacturer makes 20 chairs per day. The company receives an order for 100 chairs. Write an equation that will enable you to calculate how many days it will take the company to make all 100 chairs.

4. Avery drives 20 miles to visit his uncle. It takes him exactly 30 minutes to make the trip. Write an equation that shows what speed Avery was driving, assuming he drove the same speed throughout the trip.

5. An art teacher cuts a piece of construction paper in the shape of a rectangle. The rectangle has an area of 22 square inches. If the length of the rectangle is 11 inches, write an equation that will enable you to find the width of the rectangle.

6. Colin earns $8.00 per hour working at the grocery store. He works 4 hours on Friday and 6 hours on Saturday. Write an equation that shows the total amount of money Colin earned on those two days.

7. Alexandra spends 1 hour and 45 minutes at the pool. She then spends 30 minutes at the store. Write an equation that shows the total amount of hours Alexandra spends at the pool and at the store.

10. Mr. Jensen obtained movie tickets for his family at a discount. The regular price of the tickets was $10.00 each. The discount price was 25% off the regular price. If Mr. Jensen purchased 6 tickets, write an equation that shows the total amount Mr. Jensen paid.

8. Sylvie is making a chocolate cake recipe that calls for 2 cups of sugar, $1\frac{3}{4}$ cups of flour, and $\frac{3}{4}$ cup of cocoa powder. She plans to double the recipe to make two cakes. Write an equation that shows how many total ounces of sugar, flour, and cocoa powder she will need for the two cakes.

11. The Montoya family is expanding their vegetable garden. The distance around their original garden is 26 feet. The new garden will be twice as long and three times as wide as the original garden. The total distance around the new garden will be 64 feet. What are the length and width of the original garden?

9. Micah mows lawns in his neighborhood as a summer job. It takes him 2 hours to mow each lawn he takes care of. Write an equation that shows how many lawns Micah mows in a weekend, if he works for 8 hours per day on both days.

12. Vincenzo's Catering Company catered a business luncheon. The main course was a choice of lobster tail or tuna steak. The ratio of lobster tails to tuna steaks served was 7:2. The total number of lobster tails served was 49. How many tuna steaks were served?

13. Yvonne and Zach are twins. When they received their report cards, Yvonne had 3 more A's than Zach did. Together they had a total of 15 A's. What was the ratio of Yvonne's A's to Zach's A's?

14. A fruit smoothie costs $3.99 more than a milkshake. If an order of three fruit smoothies and seven milkshakes costs $31.87 total, how much do milkshakes and fruit smoothies cost each?

15. The dollar amounts of Jenny's paycheck and Katie's paycheck are two consecutive integers. The product of their paychecks is $1,122. Jenny's paycheck is $1 larger than Katie's. What is the amount of each woman's paycheck?

16. Two jets approach a set midpoint from locations that are 2,800 miles apart. The first jet flies at a speed of 200 miles per hour, and the second jet flies at a speed of 350 miles per hour. The jets arrive at the midpoint point at the same time. If the first plane left at 10:00 a.m., what time did the second plane depart?

17. Kim purchased 23 shirts for a total of $281.21. All of the shirts had an original price of $14.95, but some had been reduced to $7.99. How many did she buy at each price?

18. Alfonse receives a check for the dividends he has earned on two stocks. The first stock has yielded a dividend of $27. The second stock has yielded a $24 dividend. The total dividends earned are 1.2 percent of the original investment in the stocks. What was the amount of the original investment?

19. There are three sets of floor tiles for sale at an auction. The largest set contains 90 tiles. The two smaller sets contain tiles in a ratio of 2:1. There are a total of 180 tiles between the three sets. How many tiles are in each of the smaller two sets?

20. A local art gallery needs to have a new painting framed for display. The new painting is rectangular, and its length is two times its width. The painting has a total area of 288 square inches. Find the length and width, in feet, of the painting.

21. If three times the length of a race track is increased by 8 miles, the result is 20 miles less than the square of the track's original length. Find the length, in miles, of the race track.

22. Albert has one pound of a pesticide mixture. It contains 69 percent of a compound called diatomaceous earth. The remaining portion of the pesticide mixture is made up of other organic matter. How many ounces of other organic matter does the mixture contain?

23. Mandy is 21 months older than her sister Amy, and the sum of their ages is 37 months. How old is Mandy, in years and months?

24. Craig has $9.75 in nickels and dimes. If he has exactly twice as many nickels as dimes, how many of each type of coin does he have?

25. Mr. Carter makes two kinds of cheese, sharp cheddar and mild cheddar. The sharp cheddar he sells is five times as aged as the mild cheddar. If both cheeses remain aging in the cooler for another seven months, the sharp cheddar will be exactly three times as aged as the mild cheddar. For how many months have each of the cheeses been aging?

Practice Question Solutions

Setting up word problems

1. One equation that can be used to set up this problem is $c = 17 - 2 + 8$.

   Let c represent the number of baseball cards that Marek has. He starts with 17 cards and gives away 2. Express this as an equation:

   $c = 17 - 2$

   Marek then adds 8 cards to this amount. Add 8 to the right side of the equation:

   $c = 17 - 2 + 8$

   This equation is set up and ready to solve.

2. One equation that can be used to set up this problem is $\frac{12}{1} = \frac{b}{4}$.

   We are told that the ice cream truck sells 12 ice cream bars per hour. Set this up as a ratio:

   $$\frac{12 \text{ ice cream bars}}{1 \text{ hour}}$$

   We are asked to find the number of ice cream bars the truck sells in 4 hours. Let b represent the missing amount:

   $$\frac{b \text{ ice cream bars}}{4 \text{ hours}}$$

   Set these two ratios equal to each other to find the value of b:

$$\frac{12 \text{ ice cream bars}}{1 \text{ hour}} = \frac{b \text{ ice cream bars}}{4 \text{ hours}}$$

$$\frac{12}{1} = \frac{b}{4}$$

This proportion is set up and ready to solve.

3. One equation that can be used to set up this problem is $d = 3 \times 0.25$.

   To set up this problem, use the distance formula:

   $d = s \times t$

   Marta walks at a speed of 3 miles per hour. Substitute 3 for her speed:

   $d = (3) \times t$

   She takes 15 minutes to get to school. Convert 15 minutes to hours: 15 minutes is one-fourth of an hour, or 0.25 hours. Substitute 0.25 into the equation for t:

   $d = (3) \times (0.25)$

   This equation is set up and ready to solve.

4. One system of equations that can be used to set up this problem is

   $$\begin{cases} j = 0.375 \times 24 \\ j + r = 24 \end{cases}$$

   Let j represent the number of games that James wins. Let r represent the number of games that Ryan wins.

   Write an equation for the number of games that James wins. The number of games is 37.5 percent of the total, or 37.5 percent of 24. This can be expressed as 0.375×24:

   $j = 0.375 \times 24$

   Write an equation for the total number of games played:

   $j + r = 24$

This gives us a system of two equations:

$$\begin{cases} j = 0.375 \times 24 \\ j + r = 24 \end{cases}$$

We can use this system of equations to determine the number of games, r, that Ryan wins. This problem is set up and ready to solve.

This problem is ready to solve!

5. One equation that can be used to set up this problem is $3(1.99) + 2(b) = 10.11$.

Let b represent the cost of each loaf of bread. Write an equation for the total purchase made by Mrs. Lopez:

(3 gallons of milk × \$1.99 per gallon) + (2 loaves of bread × b) = \$10.11

This can be shortened as follows:

$3(1.99) + 2(b) = 10.11$

This word problem is now set up, and we're ready to solve for b.

6. One equation that can be used to set up this problem is $A = 9 \times 13.5$.

Let A represent the area of Owen's room. Owen's room is in the shape of a rectangle, so we start with the formula for the area of a rectangle: $A = $ length × width. Substitute 9 feet for the length and 13.5 feet for the width in the equation:

$A = l \times w$
$A = (9) \times (13.5)$

7. One equation that can be used to set up this problem is $3.5 = 2(0.25 + t)$.

Write an equation that shows the relationship between the time Kelly spent at the library and the time she spent walking home and taking her dog to the park:

Time spent at library = 2 × (time spent walking home + time spent taking the dog to the park)

Kelly spent 3.5 hours at the library. Substitute 3.5 hours for the time spent at the library:

3.5 hours = 2 × (time spent walking home + time spent taking the dog to the park)

Kelly spends 15 minutes walking home. Be careful! The question asks for a time given in hours. So, we must convert minutes to hours before substituting in the equation: 15 minutes equals one-quarter hour, or 0.25 hours.

3.5 hours = 2 × (0.25 hours + time spent taking the dog to the park)

Let t represent the time Kelly spent taking her dog to the park. Substitute t into the equation:

3.5 hours = 2 × (0.25 hours + t)

The equation can be shortened as shown:

$3.5 = 2(0.25 + t)$

This problem is now ready to solve.

8. One equation that can be used to set up this problem is $w = \dfrac{23 + 34 + 17}{16}$.

   The weights of the packages are given in ounces, but the question asks for a weight in pounds. To set up this problem, we will have to convert ounces to pounds.

   Let w represent the total weight of the packages. Set up an equation for the total weight of the packages:

   $w = 23$ ounces $+ 34$ ounces $+ 17$ ounces

   There are 16 ounces in 1 pound. To convert ounces to pounds, we must divide the total number of ounces by 16:

   $w = \dfrac{23 + 34 + 17}{16}$

   This problem is set up and ready to solve.

9. One equation that can be used to set up this problem is $\dfrac{1}{5} = \dfrac{a}{75}$.

   To set up this problem, we must create a proportion. The squirrel stores 1 acorn every five minutes. Write this as a fraction:

   $\dfrac{1 \text{ acorn}}{5 \text{ minutes}}$

   We are asked to find the number of acorns that could be stored in one hour and 15 minutes, or 75 total minutes. Let a represent the number of acorns stored in 75 minutes:

   $\dfrac{a \text{ acorns}}{75 \text{ minutes}}$

   Set these two fractions equal to each other to create a proportion:

   $\dfrac{1 \text{ acorn}}{5 \text{ minutes}} = \dfrac{a \text{ acorns}}{75 \text{ minutes}}$

   $\dfrac{1}{5} = \dfrac{a}{75}$

   This problem is set up and ready to solve.

10. One system of equations that can be used to set up this problem is:

$$\begin{cases} 200a = 400b \\ a + b = 7.5 \end{cases}$$

Using the distance formula, let a represent the time it took for the jet to travel from City A to City B:

$d = 200 \times a$

Let b represent the time it took for the jet to travel from City B to City A:

$d = 400 \times b$

The distance traveled on both trips is the same, so these two equations can be set equal to one another:

$200 \times a = 400 \times b$

Write an equation for the total time it took to make the trip:

$a + b = 7.5$

This gives us the following system of equations: $\begin{cases} 200a = 400b \\ a + b = 7.5 \end{cases}$

Solving for a and b will allow us to find the time traveled on each trip. We can the substitute one of those times into the distance formula to find the distance between the cities.

Solving word problems

11. The correct answer is 8 minutes for Amber and 80 minutes for Jesse.

Let the variable a represent the amount of time it took Amber to do her homework. Let j represent the amount of time it took Jesse to do his homework. Create two equations from the information given:

$j = 72 + a$

$j = 64 + 2a$

Both of these equations represent the number of minutes it took Jesse to do his homework. The two equations represent the same amount of time, so we can set these two equations equal to one another:

$$72 + a = 64 + 2a$$
$$72 - 64 + a = 64 - 64 + 2a$$
$$72 - 64 + a = 2a$$
$$8 + a = 2a$$
$$8 + a - a = 2a - a$$
$$8 = 2a - a$$
$$a = 8$$

Amber spent 8 minutes finishing her homework. Substitute 8 for a in either of the equations above:

$j = 72 + a$

$j = 72 + (8)$

$j = 80$

It took 8 minutes for Amber to complete her homework and 80 minutes for Jesse to complete his.

12. The correct answer is $29.00.

The purchase price is 25 percent off of the regular price. Let p represent the original purchase price. The term $0.25p$ represents the 25 percent discount. Set up an equation:

$p - 0.25p = \$21.75$

This lets us solve the equation for the original purchase price, p:

$$p - 0.25p = 21.75$$
$$0.75p = 21.75$$
$$\frac{0.75p}{0.75} = \frac{21.75}{0.75}$$
$$p = \frac{21.75}{0.75}$$
$$p = 29.00$$

The original purchase price was $29.00.

13. The correct answer is 60 kilometers.

The formula for distance is speed multiplied by time:

Distance = speed × time

Let d represent the total distance of Jake's training route. Set up an equation using the information given:

$d = 30 \times 2$

$d = 60$

The length of the training route should be 60 kilometers.

14. The correct answer is 180 mph for the first train and 324 mph for the second train.

Let a represent the speed of the first train. Use the variable b to represent the speed of the second train.

The formula for distance is speed multiplied by time:

Distance = speed × time

In this problem, both trains travel the same distance. Therefore, we can create expressions for the distance traveled by each train and set these expressions equal to one another.

Train 1 reached the passing point in 9 hours. The distance traveled by Train 1 can be represented by its speed times the amount of time it took, or $a \times 9$. Train 2 reached the passing point in 5 hours. The distance traveled by Train 2 can be represented by $b \times 5$. Create an equation:

$$a \times 9 = b \times 5$$
$$9a = 5b$$

We also know that Train 2 was traveling 144 mph faster than Train 1. We can create a second equation to represent this relationship:

$$b = a + 144$$

Substitute $a + 144$ for b in the first equation:

$$9a = 5b$$
$$9a = 5(a + 144)$$
$$9a = 5a + 720$$
$$9a - 5a = 5a - 5a + 720$$
$$4a = 5a - 5a + 720$$
$$4a = 720$$
$$\frac{4a}{4} = \frac{720}{4}$$
$$a = \frac{720}{4}$$
$$a = 180$$

The speed of the first train was 180 miles per hour. Substitute 180 for a in either equation:

$$a + 144 = b$$
$$(180) + 144 = b$$
$$b = 324$$

The speed of Train 1 was 180 miles per hour, and the speed of Train 2 was 324 miles per hour.

15. The correct answer is $777.27 in the standard savings account and $362.73 in the CD.

Let s represent the investment in the standard savings account. Let c represent the investment in the CD. Then, create an equation for the initial investment:

$$s + c = 1,140$$

We are told that the standard savings account paid out 1.4 percent interest, while the CD paid out 3 percent annual interest. Let $0.014s$ represent the amount of interest earned by the savings account in a year, and $0.03c$ represent the interested earned by the CD. The accounts both paid out the same amount of interest, so the expressions for interest earned can be set equal to one another:

$$0.014s = 0.03c$$

This gives us a system of equations. To solve, begin by solving the first equation for c:

$$s + c = 1,140$$
$$s - s + c = 1,140 - s$$
$$c = 1,140 - s$$

Substitute $1,140 - s$ for c in the second equation:

$$0.014s = 0.03c$$
$$0.014s = 0.03(1,140 - s)$$
$$0.014s = 34.2 - 0.03s$$
$$0.014s + 0.03s = 34.2 - 0.03s + 0.03s$$
$$0.044s = 34.2 - 0.03s + 0.03s$$
$$0.044s = 34.2$$
$$\frac{0.044s}{0.044} = \frac{34.2}{0.044}$$
$$s = \frac{34.2}{0.044}$$
$$s \approx 777.27$$

The value of s is approximately 777.27.

The $\approx$ symbol in the equation means "approximately."

Since the figures must be given in dollars and cents, we can round the answer to 777.27. Substitute 777.27 for s in the first equation, and solve for c:

$$s + c = 1,140$$
$$777.27 + c = 1,140$$
$$777.27 - 777.27 + c = 1,140 - 777.27$$
$$c = 1,140 - 777.27$$
$$c = 362.73$$

Therefore, $777.27 was invested in the savings account, and $362.73 was invested in the CD.

16. The correct answer is 67 candles, for a total of $268.00.

Let c represent the number of candles sold. Write an equation for the number of candles that Anna needs to sell:

$$62 + (4c) = 327$$

Then solve for c:

$$62 + (4c) = 327$$
$$62 - 62 + 4c = 327 - 62$$
$$4c = 327 - 62$$
$$4c = 265$$
$$\frac{4c}{4} = \frac{265}{4}$$
$$c = \frac{265}{4}$$
$$c = 66.25$$

Rounding up to the nearest whole number, Anna must sell 67 candles for a total of $268.00 in order to earn enough money to make her car payment.

17. The correct answer is 11 months.

Let m represent the number of months that it will take for Drake to master all of the required moves. Then write an equation:

$$5m + 21 = 76$$

Next, solve for m:

$$5m + 21 = 76$$
$$5m + 21 - 21 = 76 - 21$$
$$5m = 76 - 21$$
$$5m = 55$$
$$\frac{5m}{5} = \frac{55}{5}$$
$$m = \frac{55}{5}$$
$$m = 11$$

It will take Drake 11 months to earn his brown belt.

18. The correct answer is $19.00 on Oil Slick and $23.00 on Sweet Tea.

Let x represent the amount placed on Oil Slick and y represent the amount placed on Sweet Tea. The total original bet was $42.00, so we can write an equation:

$x + y = 42$

Next, write an equation for the total winnings:

$4x + 2y = 122$

Solve the first equation for y:

$$x + y = 42$$
$$x + y - x = 42 - x$$
$$y = 42 - x$$

Then substitute $42 - x$ for y in the second equation:

$$4x + 2y = 122$$
$$4x + 2(42 - x) = 122$$
$$4x + 84 - 2x = 122$$
$$4x - 2x + 84 = 122$$
$$2x + 84 = 122$$
$$2x + 84 - 84 = 122 - 84$$
$$2x = 122 - 84$$
$$2x = 38$$
$$\frac{2x}{2} = \frac{38}{2}$$
$$x = \frac{38}{2}$$
$$x = 19$$

Substitute 19 for x in the first equation, and solve for y:

$$x + y = 42$$
$$19 + y = 42$$
$$19 - 19 + y = 42 - 19$$
$$y = 42 - 19$$
$$y = 23$$

The woman bet $19.00 on Oil Slick and $23.00 on Sweet Tea.

19. The correct answer is Level 3.

Let l represent the number of levels that could be acheived with Brendan's points. He has earned a total of 2,468 points,

and each new level requires 720 points. Write an equation:

$$l = \frac{2,468}{720}$$

Then solve for l:

$$l = \frac{2,468}{720}$$
$$l \approx 3.43$$

The result is approximately 3.43 levels. Brendan does not yet have enough points to reach Level 4. When he ended the game Saturday, he was at Level 3.

20. The correct answer is 87 hours and 30 minutes.

First, write an equation to determine how many hours Ashley needs to work to save up enough money for the computer. Let h represent the number of hours she must work:

$$190 + 8.00h = 890$$

Then solve the equation for h:

$$190 + 8.00h = 890$$
$$190 - 190 + 8.00h = 890 - 190$$
$$8.00h = 890 - 190$$
$$8.00h = 700$$
$$8.00h = 700$$
$$\frac{8.00h}{8.00} = \frac{700}{8.00}$$
$$h = \frac{700}{8.00}$$
$$h = 87.5$$

Ashley must work 87.5 hours. Converted into hours and minutes, this is 87 hours and 30 minutes.

Chapter Review Solutions

$$\frac{20 \text{ chairs}}{1 \text{ day}}$$

Let d represent the number of days needed to make 100 chairs. Write this as a fraction:

$$\frac{100 \text{ chairs}}{d \text{ days}}$$

Set the two fractions equal to each other to form a proportion:

$$\frac{20}{1} = \frac{100}{d}$$

1. One equation that can be used to set up this problem is $j = 17 + 4 + 12$.

   Let j represent Janet's age. Stuart is 17 years old, and Robin is 4 years older than Stuart. Robin's age is equal to $17 + 4$. Janet is 12 years older than Robin, so Janet's age can be written as $j = 17 + 4 + 12$.

2. One equation that can be used to set up this problem is $s = 17 - 0.30(17)$.

   Let s represent the sale price that Mr. Spector paid for the kit. The sale price, s, is equal to the regular price minus the discount. Substitute \$17.00 in the equation for the regular price:

   $s = \$17.00 -$ amount of discount

   The discount is 30% off of the regular price. Substitute $0.30 \times \$17.00$ for the amount of the discount:

   $s = \$17.00 - (0.30 \times \$17.00)$

   The equation can be shortened even further as shown:

   $s = 17 - 0.30(17)$

3. One equation that can be used to set up this problem is $\frac{20}{1} = \frac{100}{d}$.

   The furniture company makes 20 chairs in one day. Write this rate as a fraction:

4. One equation that can be used to set up this problem is $0.5s = 20$.

   Start with the distance formula: $d = s \times t$. Substitute 20 miles for the distance traveled and 30 minutes for the time. Since we are looking for speed in terms of miles per hour, we must convert minutes to hours first. There are 60 minutes in an hour, so 30 minutes equals 0.5 hours:

   $$d = s \times t$$
   $$(20) = s \times (0.5)$$

   Move the s to the right side of the equation:

   $$(20) = s \times (0.5)$$
   $$20 = 0.5s$$
   $$0.5s = 20$$

   This equation can be used to calculate the speed at which Avery was driving.

5. One equation that can be used to set up this problem is $22 = 11 \times w$.

   Start with the formula for the area of a rectangle: $A = l \times w$. Substitute 22 for A and 11 for l into the equation:

$$A = l \times w$$
$$(22) = (11) \times w$$

This equation will enable us to find the width of the rectangle.

6. One equation that can be used to set up this problem is $t = \$8.00(4 + 6)$.

    Let t represent the total amount of money Colin earned. The amount of money earned in one day is equal to the number of hours worked times the amount earned per hour. Colin worked 4 hours on Friday at $8.00 per hour. This can be represented as $\$8.00 \times 4$. He worked 6 hours on Saturday at $8.00 per hour. This can be represented as $\$8.00 \times 6$. Write an equation that reflects the earnings from both days:

    $$t = (\$8.00 \times 4) + (\$8.00 \times 6)$$

    This can also be written with the $8.00 factored out:

    $$t = \$8.00(4 + 6)$$

7. One equation that can be used to set up this problem is $h = 1.75 + 0.5$.

    The question asks for an equation written in terms of hours. Some of the information is given in minutes, so we must convert minutes to hours first.

    Alexandra spent 1 hour and 45 minutes at the pool. This is equivalent to 1.75 hours, since 45 minutes is equal to three-quarters of an hour, or 0.75 hours. She also spent 30 minutes, or 0.5 hours, at the store. Let h represent the total amount of hours Alexandra spent at the pool and at the store:

    $$h = 1.75 + 0.5$$

8. One equation that can be used to set up this problem is $z = 8 \times \left[2\left(2 + 1\frac{3}{4} + \frac{3}{4}\right)\right]$.

Let z represent the total number of ounces of these ingredients needed for the two cakes. First, add together the total number of cups of ingredients needed to make one cake:

$$\text{cups for one cake} = 2 + 1\frac{3}{4} + \frac{3}{4}$$

Multiply this amount by 2, for both cakes:

$$\text{cups for both cakes} = 2\left(2 + 1\frac{3}{4} + \frac{3}{4}\right)$$

The question asks for the number of ounces needed for the two cakes, so convert the cups to ounces. There are 8 ounces in a cup. Multiply the right side of the equation by 8:

$$\text{ounces for both cakes} = 8 \times \left[2\left(2 + 1\frac{3}{4} + \frac{3}{4}\right)\right]$$

Substitute z for the total number of ounces of ingredients needed:

$$z = 8 \times \left[2\left(2 + 1\frac{3}{4} + \frac{3}{4}\right)\right]$$

9. One equation that can be used to set up this problem is $\frac{1}{2} = \frac{n}{16}$.

    Micah mows 1 lawn in 2 hours. Write this as a fraction:

    $$\frac{1 \text{ lawn}}{2 \text{ hours}}$$

    Let n represent the total number of lawns Micah mows in both days of the weekend. He works for 8 hours per day on both days, or a total of 16 hours. Write this as a fraction:

    $$\frac{n \text{ lawns}}{16 \text{ hours}}$$

    Set the two fractions equal to each other:

    $$\frac{1 \text{ lawn}}{2 \text{ hours}} = \frac{n \text{ lawns}}{16 \text{ hours}}$$
    $$\frac{1}{2} = \frac{n}{16}$$

10. One equation that can be used to set up this problem is $t = 6 \times [10 - 0.25(10)]$.

Let t represent the total amount Mr. Jensen paid for the discounted tickets. The discounted price is equal to the regular price minus the amount of the discount. Substitute $10.00 in the equation for the regular price:

discounted price = $10.00 – amount of discount

The discount is 25% off of the regular price. Substitute 0.25 × $10.00 for the amount of the discount:

discounted price = $10.00 – (0.25 × $10.00)

The equation can be shortened even further as shown:

discounted price = $10 - 0.25(10)$

Mr. Jensen purchased 6 tickets total, so multiply the right side of the equation by 6:

$$t = 6 \times [10 - 0.25(10)]$$

This equation allows us to calculate t, the total amount Mr. Jensen paid.

11. The correct answer is 7 feet long and 6 feet wide.

Let l represent the length and w represent the width of the original garden. Write an equation for the distance around the original garden:

$2l + 2w = 26$

Then write an equation for the distance around the new garden:

$2(2l) + 2(3w) = 64$

Solve the first equation for w:

$$2l + 2w = 26$$
$$2l - 2l + 2w = 26 - 2l$$
$$2w = 26 - 2l$$
$$\frac{2w}{2} = \frac{26}{2} - \frac{2l}{2}$$
$$w = \frac{26}{2} - \frac{2l}{2}$$
$$w = 13 - l$$

Substitute $13 - l$ for w in the second equation:

$$2(2l) + 2(3w) = 64$$
$$2(2l) + 2[3(13 - l)] = 64$$
$$4l + 2[3(13 - l)] = 64$$
$$4l + 2(39 - 3l) = 64$$
$$4l + 78 - 6l = 64$$
$$4l - 6l + 78 = 64$$
$$-2l + 78 = 64$$
$$-2l + 78 - 78 = 64 - 78$$
$$-2l = 64 - 78$$
$$-2l = -14$$
$$\frac{-2l}{-2} = \frac{-14}{-2}$$
$$l = \frac{-14}{-2}$$
$$l = 7$$

Then substitute 7 for l in the original equation:

$$2l + 2w = 26$$
$$2(7) + 2w = 26$$
$$14 + 2w = 26$$
$$14 - 14 + 2w = 26 - 14$$
$$2w = 26 - 14$$
$$2w = 12$$
$$\frac{2w}{2} = \frac{12}{2}$$
$$w = \frac{12}{2}$$
$$w = 6$$

The original garden has a length of 7 feet and a width of 6 feet.

12. The correct answer is 14.

We know that 49 lobster tails were served. We also know that the ratio of lobster tails to tuna steaks was 7 to 2. Using this information, we can set up a proportion to determine the number of tuna steaks served:

$$\frac{7}{2} = \frac{\text{lobster tails}}{\text{tuna steaks}}$$

$$\frac{7}{2} = \frac{49}{t}$$

In the equation, t stands for the number of tuna steaks served. Cross multiply to solve for t:

$$\frac{7}{2} = \frac{49}{t}$$
$$7 \times t = 2 \times 49$$
$$7t = 2 \times 49$$
$$7t = 98$$
$$\frac{7t}{7} = \frac{98}{7}$$
$$t = \frac{98}{7}$$
$$t = 14$$

Therefore, 14 tuna steaks were served.

13. The correct answer is 9:6, which reduces to 3:2.

Let y represent the number of A's Yvonne received, and let z represent the number of A's Zach received. Write an equation that represents the total number of A's the twins received:

$$y + z = 15$$

Then write an equation that shows that Yvonne had 3 more A's than Zach:

$$y = z + 3$$

Substitute $z + 3$ for y in the first equation and solve:

$$y + z = 15$$
$$(z + 3) + z = 15$$
$$z + z + 3 = 15$$
$$2z + 3 = 15$$
$$2z + 3 + 3 = 15 - 3$$
$$2z = 15 - 3$$
$$2z = 12$$
$$\frac{2z}{2} = \frac{12}{2}$$
$$z = \frac{12}{2}$$
$$z = 6$$

Zach had a total of 6 A's. Substitute 6 in for z in either equation and solve:

$$y + z = 15$$
$$y + (6) = 15$$
$$y + 6 - 6 = 15 - 6$$
$$y = 15 - 6$$
$$y = 9$$

Yvonne had a total of 9 A's. After finding both y and z, set these numbers up in a ratio. In this case, $y = 9$ and $z = 6$, so the ratio of Yvonne's A's to Zach's A's is 9:6. Both of those terms are divisible by 3, so the ratio can be simplified.

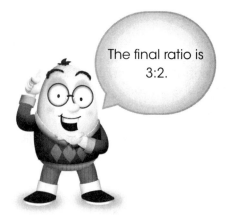

The final ratio is 3:2.

14. The correct answer is $1.99 for a milkshake and $5.98 for a fruit smoothie.

Let m represent the cost of a milkshake and f represent the cost of a fruit smoothie. Set up an equation for the total cost of the order:

$7m + 3f = \$31.87$

Then, set up an equation for the cost of a fruit smoothie:

$m + 3.99 = f$

Substitute $m + 3.99$ in for f in the first equation, and solve:

$$7s + 3f = 31.87$$
$$7m + 3(3.99 + m) = 31.87$$
$$7m + 11.97 + 3m = 31.87$$
$$10m + 11.97 = 31.87$$
$$10m + 11.97 - 11.97 = 31.87 - 11.97$$
$$10m = 19.90$$
$$\frac{10m}{10} = \frac{19.90}{10}$$
$$m = \frac{19.90}{10}$$
$$m = 1.99$$

Substitute 1.99 in for m in either equation and solve for f.

$$m + 3.99 = f$$
$$(1.99) + 3.99 = f$$
$$5.98 = f$$

The price of a milkshake is $1.99 and the price of a fruit smoothie is $5.98.

15. The correct answers are $33.00 and $34.00.

Let n represent the amount of the smaller paycheck. The larger paycheck is equal to $n + 1$. Write an equation for the product of the two paychecks:

$n \times (n + 1) = 1,122$

Then reorganize the equation to form a quadratic equation:

$$n \times (n + 1) = 1,122$$
$$n^2 + n = 1,122$$
$$n^2 + n - 1,122 = 0$$

Factor the equation using the reverse FOIL method:

$$n^2 + n - 1,122 = 0$$
$$(n + 34)(n - 33) = 0$$

Set each factor equal to 0 and solve:

$$n + 34 = 0 \qquad\qquad n - 33 = 0$$
$$n + 34 - 34 = 0 - 34 \qquad n - 33 + 33 = 0 + 33$$
$$n = 0 - 34 \qquad\qquad n = 0 + 33$$
$$n = -34 \qquad\qquad\qquad n = 33$$

Since the answers refer to dollar amounts, the negative number can be eliminated. However, we know that the larger paycheck is one dollar more than the smaller paycheck, n. Add 1 to n for the amount of the larger check:

$$n + 1 = (33) + 1$$
$$n + 1 = 34$$

The values of the paychecks are $33.00 and $34.00.

16.

The correct answer is 1:00 p.m.

Let x represent the amount of time traveled by the first jet and y represent the amount of time traveled by the second jet. The formula for calculating distance is:

Distance = speed $\times$ time

We can write this in shorthand as $D = s \times t$. The formula for calculating time is therefore time = distance ÷ speed, or $t = \dfrac{D}{s}$.

Use this formula to determine the time it takes each jet to reach its destination. The jets meet at a midpoint between two locations that are 2,800 miles apart, so the distance traveled by each jet is 1,400 miles:

$$x = \frac{D}{s} \qquad\qquad y = \frac{D}{s}$$

$$x = \frac{1,400 \text{ miles}}{200 \text{ mph}} \qquad x = \frac{1,400 \text{ miles}}{350 \text{ mph}}$$

$$x = 7 \text{ hours} \qquad\qquad x = 4 \text{ hours}$$

The first jet traveled for 7 hours, and the second jet traveled for 4 hours. The first jet left at 10:00 a.m., so it arrived at 5:00 p.m. The second jet also arrived at 5:00 p.m., and it departed 4 hours earlier. The second jet departed at 1:00 p.m.

17. The correct answer is 14 shirts at $14.95 each and 9 shirts at $7.99 each.

Let x represent the number of shirts that cost $14.95 each. Let y represent the number of shirts that cost $7.99 each. Create an equation that represents the total purchase:

$$14.95x + 7.99y = 281.21$$

Create a second equation that reflects the total number of shirts:

$$x + y = 23$$

The second equation is easier to work with, so solve the second equation for y:

$$x + y = 23$$
$$x - x + y = 23 - x$$
$$y = 23 - x$$

Substitute $23 - x$ for y in the first equation, and solve for x:

$$14.95x + 7.99y = 281.21$$
$$14.95x + 7.99(23 - x) = 281.21$$
$$14.95x + 183.77 - 7.99x = 281.21$$
$$14.95x - 7.99x + 183.77 = 281.21$$
$$6.96x + 183.77 = 281.21$$
$$6.96x + 183.77 - 183.77 = 281.21 - 183.77$$
$$6.96x = 281.21 - 183.77$$
$$6.96x = 97.44$$
$$\frac{6.96x}{6.96} = \frac{97.44}{6.96}$$
$$x = \frac{97.44}{6.96}$$
$$x = 14$$

Substitute 14 in for x in either equation and solve for y:

$$x + y = 23$$
$$(14) + y = 23$$
$$14 - 14 + y = 23 - 14$$
$$y = 23 - 14$$
$$y = 9$$

The correct answer is 14 shirts at $14.95 and 9 shirts at $7.99.

18. The correct answer is $4,250.

First, create an equation for the total dividend payment. Let d represent the total dividend payment:

$$27 + 24 = d$$

Let n represent the amount of the initial investment. The dividend payment is 1.2 percent of the initial investment. To find the dividend payment d, we multiply the initial investment by 1.2 percent, or 0.012:

$$d = 0.012n$$

Set these two equations equal to one another, and solve for n:

$$27 + 24 = 0.012n$$
$$51 = 0.012n$$
$$0.012n = 51$$
$$\frac{0.012n}{0.012} = \frac{51}{0.012}$$
$$n = \frac{51}{0.012}$$
$$n = 4,250$$

The amount of the initial investment was $4,250.

19. The correct answer is 60 tiles in one of the smaller sets and 30 tiles in the other smaller set.

Let a represent the number of tiles in the first small set, and b represent the number of tiles in the second small set.

Create an equation for the total number of tiles:

$$90 + a + b = 180$$

Then, write an equation reflecting the ratio of a to b:

$$\frac{a}{b} = \frac{2}{1}$$
$$a = 2b$$

Substitute $2b$ for a in the first equation and solve for b:

$$90 + a + b = 180$$
$$90 + (2b) + b = 180$$
$$90 + 3b = 180$$
$$90 - 90 + 3b = 180 - 90$$
$$3b = 180 - 90$$
$$3b = 90$$
$$\frac{3b}{3} = \frac{90}{3}$$
$$b = \frac{90}{3}$$
$$b = 30$$

Substitute 30 for b in either equation and solve for a:

$$90 + a + b = 180$$
$$90 + a + (30) = 180$$
$$90 + 30 + a = 180$$
$$120 + a = 180$$
$$120 - 120 + a = 180 - 120$$
$$a = 180 - 120$$
$$a = 60$$

The value of a is 60, and the value of b is 30. Therefore, there are 60 tiles in the first of the two smaller sets and 30 tiles in the second smaller set.

20. The correct answer is 2 feet long and 1 foot wide.

The formula for the area of a rectangle is *Area* = length × width, or $A = l \times w$.

Write an equation for the area of the painting:

$lw = 288$

Then write an equation for the length, l, in terms of the width, w:

$l = 2w$

Substitute $2w$ for l in the first equation, and solve for w:

$$lw = 288$$
$$(2w)w = 288$$
$$2w^2 = 288$$
$$\frac{2w^2}{2} = \frac{288}{2}$$
$$w^2 = 144$$
$$w = \sqrt{144}$$
$$w = 12$$

Substitute 12 for w in either equation and solve for l:

$l = 2w$
$l = 2(12)$
$l = 24$

The painting has a length of 24 inches and a width of 12 inches. Divide each side by 12 to convert inches into feet. The painting is is 2 feet long and 1 foot wide.

21. The correct answer is 7 miles.

First, write out an equation for the length of the track, letting t represent the length of the track. From the information provided, we can create two expressions and set them equal to each other. On the left side, put the first part of the description: "three times the length of the track is increased by 8 miles." This can be written as $3t + 8$. On the right side of the equal sign, put the second part of the description: "the result is 20 miles less than the square of the track's original length." This can be written as $(t \times t) - 20$, or $t^2 - 20$:

$3t + 8 = t^2 - 20$

Then, rewrite the equation as a quadratic equation:

$$3t + 8 = t^2 - 20$$
$$3t - 3t + 8 = t^2 - 3t - 20$$
$$8 = t^2 - 3t - 20$$
$$8 - 8 = t^2 - 3t - 20 - 8$$
$$0 = t^2 - 3t - 20 - 8$$
$$0 = t^2 - 3t - 28$$
$$t^2 - 3t - 28 = 0$$

Factor the equation using the reverse FOIL method:

$$t^2 - 3t - 28 = 0$$
$$(t + 4)(t - 7) = 0$$

Set each factor equal to 0 and solve:

$(t + 4) = 0$	$(t - 7) = 0$
$t + 4 - 4 = 0 - 4$	$t - 7 + 7 = 0 + 7$
$t = 0 - 4$	$t = 0 + 7$
$t = -4$	$t = 7$

Since the problem refers to the length of a track, the negative solution can be eliminated. The length of the track is 7 miles.

22. The correct answer is 4.96 ounces.

The pesticide mixture weighs 1 pound, or 16 ounces. It contains 69 percent diatomaceous earth. Let d represent the number of ounces of diatomaceous earth in the mixture. Set up an equation that reflects the amount of diatomaceous earth:

$d = 16 \times 0.69$

Solving for d, we see that the mixture contains 11.04 ounces of diatomaceous earth.

The remaining portion of the mixture is made up of organic matter. Subtract 11.04 ounces from 16 ounces to determine the number of ounces of organic matter in the pesticide:

$16 - 11.04 = 4.96$

23. The correct answer is 2 years and 5 months.

Let m represent Mandy's age and a represent Amy's age. Create an equation for Mandy's age:

$m = a + 21$

Then, create an equation for Mandy and Amy's combined ages:

$m + a = 37$

Substitute $a + 21$ for m in the second equation and solve for a:

$$m + a = 37$$
$$(a + 21) + a = 37$$
$$a + a + 21 = 37$$
$$2a + 21 = 37$$
$$2a + 21 - 21 = 37 - 21$$
$$2a = 37 - 21$$
$$2a = 16$$
$$\frac{2a}{2} = \frac{16}{2}$$
$$a = \frac{16}{2}$$
$$a = 8$$

Then substitute 8 for a in either equation and solve for Mandy's age:

$$a + 21 = m$$
$$(8) + 21 = m$$
$$m = 29$$

Convert the 29 months to years and months by dividing by 12. This gives a final answer of 2 years and 5 months.

24.

Let n represent the number of nickels and d represent the number of dimes.

First, write out an equation for the total amount of change:

$0.05n + 0.10d = \$9.75$

Then write an equation to represent the relationship between nickels and dimes:

$2n = d$

Substitute $2n$ for d in the first equation and solve:

$$0.05n + 0.10d = 9.75$$
$$0.05n + 0.10(2n) = 9.75$$
$$.05n + 0.20n = 9.75$$
$$0.25n = 9.75$$
$$\frac{0.25n}{0.25} = \frac{9.75}{0.25}$$
$$n = \frac{9.75}{0.25}$$
$$n = 39$$

Substitute 39 for n in either equation and solve for d:

$$2n = d$$
$$2(39) = d$$
$$d = 78$$

25. The correct answer is 7 months for the mild cheddar and 35 months for the sharp cheddar.

Let s represent the age of the sharp cheddar, in months, and m represent the age of the mild cheddar. Begin by writing an equation for the ages of the cheeses:

$s = 5m$

Then write an equation for the ages of the cheeses in seven months:

$s + 7 = 3(m + 7)$

Substitute $5m$ in for s in the second equation and solve for m:

$$s + 7 = 3(m + 7)$$
$$(5m) + 7 = 3(m + 7)$$
$$5m + 7 = 3m + 21$$
$$5m - 3m + 7 = 3m - 3m + 21$$
$$2m + 7 = 3m - 3m + 21$$
$$2m + 7 = 21$$
$$2m + 7 - 7 = 21 - 7$$
$$2m = 21 - 7$$
$$2m = 14$$
$$\frac{2m}{2} = \frac{14}{2}$$
$$m = \frac{14}{2}$$
$$m = 7$$

Then substitute 7 for m in either equation to solve for s:

$$s = 5m$$
$$s = 5(7)$$
$$s = 35$$

The sharp cheddar has aged for 35 months, and the mild cheddar has aged for 7 months.